Synthesis Lectures on Visual Computing: Computer Graphics, Animation, Computational Photography and Imaging

This series presents lectures on research and development in visual computing for an audience of professional developers, researchers, and advanced students. Topics of interest include computational photography, animation, visualization, special effects, game design, image techniques, computational geometry, modeling, rendering, and others of interest to the visual computing system developer or researcher.

Simone Cammarasana · Giuseppe Patanè

Real-Time Processing of Ultrasound Images

State-of-the-Art and Future Perspectives

Springer

Simone Cammarasana
IMATI
National Research Council
Genova, Italy

Giuseppe Patanè
IMATI
National Research Council
Genova, Italy

ISSN 2469-4215 ISSN 2469-4223 (electronic)
Synthesis Lectures on Visual Computing: Computer Graphics, Animation, Computational Photography and Imaging
ISBN 978-3-032-05804-1 ISBN 978-3-032-05805-8 (eBook)
https://doi.org/10.1007/978-3-032-05805-8

This Springer imprint is published by the registered company Springer Nature Switzerland AG
The registered company address is: Gewerbestrasse 11, 6330 Cham, Switzerland

To Michela Spagnuolo, in recognition of her invaluable support and vision in advancing CNR-IMATI and PNRR RAISE activities.

Foreword

Among the various medical imaging modalities, ultrasound has always been a fundamental tool for screening, diagnosis, and follow-up. In recent years, its relevance has expanded to include support for surgery through co-registration with magnetic resonance imaging (MRI), outpatient and home-based care through portable and ultra-portable devices, and clinical diagnoses in rural or underdeveloped areas through tele-ultrasound techniques. Its diffusion is driven by several distinctive advantages: patient safety, economic sustainability, portability, rapid set-up, and the ability to acquire and visualise data in real time.

Despite these strengths, ultrasound imaging presents intrinsic challenges when compared with other clinical imaging modalities: image quality may be compromised by noise and lower spatial resolution and image formation is highly sensitive to the operator, probe characteristics, and acquisition protocol, resulting in variability across examinations. Furthermore, while standard ultrasound provides two-dimensional acquisitions, obtaining volumetric data requires more complex, costly, and less precise instrumentation than MRI or CT.

These characteristics make ultrasound an increasingly dynamic and interdisciplinary research field. Open challenges involve advanced mathematical and computer models for signal processing, compliance with clinical and industrial requirements related to manufacturing cost and real-time visualisation, and compatibility with the physical properties of probes and acquisition modalities.

This volume presents state-of-the-art methodologies for the real-time processing of 2D and 3D ultrasound images, as well as time-variant ultrasound signals. Particular attention is devoted to noise reduction and image quality enhancement techniques, with a rigorous comparison between the main algorithms, tested on international benchmark datasets, in different anatomical districts, and evaluated both through quantitative metrics and by experts in the field. Each method is presented with mathematical rigour and with a clear discussion of its applicability, strengths, and limitations within practical ultrasound practice.

The authors also discuss emerging challenges and future developments in ultrasound imaging: the role of artificial intelligence and high-performance computing architectures, integration with other acquisition modalities (such as magnetic resonance), types of data (such as semantic data from electronic health records), and post-processing procedures (such as anatomical segmentation and pathology classification). The volume highlights novel and rapidly evolving applications, including microbubble-based techniques and focused ultrasound for therapeutic interventions.

This book thus represents a valuable resource for researchers in mathematics, computer science, and engineering, providing in-depth comparative insights into modern methodologies, and for clinicians and industrial stakeholders seeking solutions responsive to real-world diagnostic needs. Moreover, the research directions presented here naturally converge with the mission of RAISE Spoke 2—"*Smart devices and technologies for personal and remote healthcare*", part of the PNRR Innovation Ecosystem RAISE—*Robotics and AI for Socio-economic Empowerment*. The advancement of intelligent, accessible, and clinically meaningful ultrasound technologies directly contributes to the development of next-generation diagnostic tools capable of transforming both conventional clinical practice and remote healthcare delivery, in full alignment with the vision and strategic objectives of RAISE Spoke 2.

Lorenzo De Michieli
Director of Technology Transfer at
the IIT—Italian Institute of Technology
Coordinator of RAISE Spoke 2
"Smart devices and technologies for
personal and remote healthcare" PNRR
Innovation Ecosystem RAISE—Robotics
and AI for Socio-economic Empowerment

Acknowledgements Simone Cammarasana and Giuseppe Patanè (CNR-IMATI) are part of the RAISE Innovation Ecosystem, funded by the European Union—NextGenerationEU and by the Ministry of University and Research (MUR), National Recovery and Resilience Plan (NRRP), Mission 4, Component 2, Investment 1.5, project "RAISE—Robotics and AI for Socio-economic Empowerment" (ECS00000035).

Competing Interests The authors have no competing interests to declare that are relevant to the content of this manuscript.

Contents

List of Figures

List of Tables

Introduction

1

In the last few decades, several techniques have been developed for the acquisition of medical images, such as magnetic resonance (MR), ultrasonography (US), elastography, positron emission tomography (PET), and computed tomography (CT), supporting physicians in diagnosing, monitoring, and treating patient diseases. US imaging is widespread among these methodologies for its properties and advantages over other imaging modalities, such as non-invasiveness, low cost, portability, and real-time acquisition. US acquisition applies high-frequency (1–20 MHz) sound waves and a pulse-echo approach with a brightness-mode (B-mode) display to generate images of soft tissues, allowing physicians to identify anatomical structures such as internal organs, muscles, and blood vessels. US acquisition is widespread across multiple medical specialities, such as diagnosis, surgery, and emergency medicine, and is applied to different anatomical districts, e.g., cardiac, musculoskeletal, abdominal, and obstetric. US probes acquire 2D/3D images and 2D videos. In particular, 2D US videos allow the physician to characterise and analyse anatomical parts that dynamically change or move over time (e.g., heart valve), while 3D images provide a volumetric view of the anatomical part to improve its visualisation and the related diagnosis.

US acquisition deals with several challenges: US images are affected by *speckle noise*, a granular pattern generated from the superimposition of high-frequency US waves, which reduces the image quality, anatomical features and boundaries recognition, and diagnostic accuracy. US images may be affected by *low-resolution*, due to low-cost probes (e.g., ultra-handheld devices), or the clinical requirement of high-frequency videos, which supports the physician in better characterising the movement of an anatomical part at the cost of a reduced image resolution. Additionally, the *manual skill and experience* of the operator can affect the quality of US images, where different probe settings, position of the patient body, or orientation generate heterogeneous images, and diagnostic outcomes may be inconsistent. Limitations in *field of view* and *ultrasound penetration* affect the acquisition of large and

S. Cammarasana and G. Patanè, *Real-Time Processing of Ultrasound Images*, Synthesis Lectures on Visual Computing: Computer Graphics, Animation, Computational Photography and Imaging, https://doi.org/10.1007/978-3-032-05805-8_1

deep anatomical areas. Finally, ultrasound acquisition requires a *real-time* processing of the image. Industrial and clinical stakeholders (e.g., US device manufacturers and clinicians) require short execution time (e.g., 15–60 Hz). Furthermore, low *memory requirements* are necessary to reduce the manufacturing cost of the US probe.

Current trends in US imaging are driven by advancements in hardware and software, artificial intelligence (AI), and clinical applications. *Image optimisation* techniques, such as *denoising*, *super-resolution*, and *edge enhancement*, improve the visual quality of the image while preserving the anatomical features and supporting the diagnosis, treatment planning, and patient monitoring by the physician. In particular, denoising reduces the speckle noise without generating artefacts (e.g., blurring), while super-resolution reconstructs non-acquired scan lines to improve the image quality for both visual and quantitative analysis. *Fusion methods* combine US images with other modalities (e.g., MRI or CT) to account for the advantages of different acquisition methodologies (e.g., real-time US with high-resolution MRI). *Standardised datasets and benchmarks* consist of annotated images and are relevant for the evaluation of the performance of US imaging algorithms and the comparison with state-of-the-art methods. A *variety of quantitative metrics*, including peak signal-to-noise ratio and structural similarity index, are applied to assess image quality, evaluating structural, local, and spectral image properties. The integration of *artificial intelligence approaches* allows the user to process large data sets from different anatomical districts for several tasks, such as image enhancement, segmentation, classification, and quantitative analysis. AI algorithms can automatically identify anatomical structures, detect abnormalities and suggest diagnoses for a preliminary screening, to reduce the time for the physician to analyse US images and improve the diagnosis, especially in areas where access to specialist radiologists is limited. *AI-driven image acquisition* supports the operator in acquiring high-quality images by guiding the probe placement and adjusting imaging parameters. *High-performance computing* (HPC) is widespread in image processing and biomedical applications to solve computational and significant data challenges, allowing users to deal with large and heterogeneous data sets through AI-based methods.

US imaging supports post-processing and quantitative analysis. Manual *segmentation* is time-consuming and highly dependent on operator expertise. Novel unsupervised methods automatically segment key anatomical structures such as organs, tumours and blood vessels with high accuracy, supporting the physician to identify the region of interest. *Morphological analysis* processes the shapes and structures of US images to understand the spatial and structural properties of anatomical structures, such as size, shape, connectivity, and topology. Morphological analysis is applied for *classification* models, such as the recognition of benign/malignant tumours from geometrical features (e.g., tumour dimension, shape, and smoothness) to support clinicians in identifying potential issues like masses, cysts or abnormal fluid accumulation. In this context, AI models can be trained to recognise heart disease, liver cirrhosis or foetal abnormalities and support the physician with preliminary screening. *Quantitative analysis* of US data measures anatomical structures and physiological parameters to provide physicians with quantitative values to support the diagnosis.

US imaging also supports physicians in several clinical contexts. The *radiologist* accounts for US images for screening, diagnosis, and follow-up to monitor the progression of diseases and the effectiveness of treatments. The *surgeon* applies US to support the visualisation during the surgical operation, for example, to monitor the position of the needle during tumour ablation or biopsy. The *point-of-care ultrasound* (POCUS) is performed in real-time and interpreted by the physician at a bedside test, e.g., for the evaluation of pulmonary conditions, the screening of abdominal aortic aneurysms in the emergency department, the diagnosis of deep vein thrombosis, where prompt identification is essential to prevent complications such as pulmonary embolism. *Wearable ultrasound* (e.g., heart patches) allows continuous monitoring of the condition of the patients, such as heart failure or chronic diseases of organs (e.g., kidney, liver). *Tele-ultrasound* exploits 5G and cloud technologies to support real-time and remote diagnosis in rural and remote areas, or where the lack of expert radiologists is relevant.

Overview and Contributions

The book provides a comprehensive review of the rapidly expanding field of real-time processing of US images, with a potential interest for researchers, clinicians, and industries. The book provides in-depth discussions on US denoising, enhancement, and super-resolution, and illustrates the wide use of the US and its underlying potential in hospitals and territorial care. We focus on the *real-time processing of 2D/3D and time-varying US images* based on

- *Denoising and enhancement for noise removal*, preservation of edges of the anatomical features and texture properties, and real-time computation without generating artefacts. Noise reduction improves the visual evaluation by the physician and the application of post-processing methods (e.g., segmentation).
- *Super-resolution* by increasing the image resolution and reconstructing non-acquired scan lines. The super-resolution of US videos with low spatial resolution and a high frequency (e.g., for the cardiac district) generates high-frequency 2D US video with an increased spatial resolution of each frame, thus overcoming the primary limits of current US probes, whose spatial resolution decreases as the acquisition frequency increases.

For the *real-time processing and visualisation* of US signals, we discuss state-of-the-art work on real-time denoising and super-resolution of US 2D/3D images and videos, which (i) take into account the high variability of US signals, (ii) are general for the underlying anatomical (e.g., musculoskeletal, abdominal, cardiac) districts, and (iii) are compliant with industrial requirements in terms of computational cost and memory requirements (e.g., on commercial US machines). Denoising and super-resolution are part of the processing methodologies for the enhanced visualisation of US images. We address the following challenges of US processing and visualisation:

- a *speckle noise* generated by the superimposition of the ultrasound waves corrupts the resulting image. It significantly affects its visualisation, resulting in a poor image contrast degree and inadequate and non-identifiable characteristics of the tissues and structure features. Speckle noise influences the accuracy of diagnosis and post-processing methods by affecting the identification of geometries and edges of the anatomical features;
- a *low-resolution, anisotropy, and variability* of the acquired data (e.g., acquisition direction, resolution, frame rate) due to the US probe characteristics and the anatomy of the input district. The spatial and temporal resolution of the US video affects the level of detail of static and dynamic anatomical properties; and
- the *computational aspects* (e.g., real-time processing, memory storage) for real-time visualisation, processing, and clinical evaluation. The real-time calculation is the primary industrial and clinical constraint, as the physician must see the changes in the signal as the probe moves in real-time.

For the analysis and comparison of state-of-the-art work, we will discuss the following criteria:

- *generality* of the denoising and super-resolution algorithms to the input images (e.g., 2D/3D images, 2D videos), resolution (e.g., low-resolution, isotropic/anisotropic images), and anatomical districts (e.g., cardiac, musculoskeletal);
- *smoothing quality, signal enhancement, and feature preservation*, including the geometries/textures of organs and tissues, and artefact reduction in terms of blurring and shading. These properties will be addressed through parameter tunings, such as soft or hard smoothing, and different up-sampling factors (e.g., 2X and 4X);
- *real-time processing* due to the industrial and clinical requirements of US image processing methods, and *computational cost and memory consumption* compatible with real-time visualisation constraints and hardware requirements (e.g., CPU and GPU computing power); and
- *quantitative comparison* based on local and structural metrics, qualitative and clinical validation in the industrial and medical domains.

As the main content of the book, we discuss:

- *Ultrasound Imaging: Overview and Challenges* We review previous work on the acquisition of US signals (i.e., 2D/3D images and 2D videos), discuss their main spatial and temporal properties, and how these properties affect the visualisation and processing of US signals. Finally, we describe US benchmarks, metrics, and tools that exploit deep learning and high-performance computing (Chap. 2).
- *Real-Time US Denoising and Enhancement* For the real-time US visualisation based on denoising and enhancement, it is relevant to address speckle noise, which is a multiplicative noise generated when the scale of the structures of the organ tissue is close to

the ultrasonic wavelengths; different waves are reflected and overlap, causing the perturbation in the signals and appearing in the output image as darker and brighter grains. Ultrasonic image denoising has pros and cons: noise reduction improves visual evaluation, and the application of post-processing methods that use local information and are therefore influenced by noisy elements that alter the accuracy of numerical operators, such as convolution and interpolation. At the same time, applying denoising algorithms changes the image and, therefore, potentially the contours and anatomical structures. Finally, image denoising increases the time from data acquisition to screen display for the physician. We review previous work on the real-time denoising of arbitrary US signals, such as 2D videos and 3D images; in the discussion, we perform a quantitative comparison through quantitative metrics, a qualitative evaluation performed by US experts on US images acquired from different anatomical (e.g., musculoskeletal, obstetric, and abdominal) districts, and an analysis of computational cost and execution time (Chaps. 3, 4, and 5).

- *Real-Time US Super-Resolution* The spatial resolution of a US system is limited by the diffraction limit to length scales of approximately half of the wavelength of the transmitted beam; two or more point reflectors that are spaced more closely than this limit cannot be distinguished. For the real-time US visualisation and enhancement based on super-resolution, it is fundamental to address the US image resolution that is affected by the required frequency of the video, since some anatomical districts (e.g., cardiac) need a high acquisition frequency to accurately acquire the behaviour of anatomical features that quickly change over time. The US image super-resolution is relevant to improving the quality of the image, its visual interpretation by the physician, and post-processing steps, e.g., classification, diagnosis, and segmentation. We review state-of-the-art work on the real-time super-resolution of US signals, focusing on super-resolution for reconstructing the non-acquired scan lines. We perform a quantitative and qualitative comparison of super-resolution methods on an extensive collection of US signals (i.e., 2D/3D images, 2D videos) acquired from different anatomical districts (Chaps. 6 and 7).
- *Future Technological and Research Perspectives* We discuss future work in terms of analysis and development of novel methods for the further processing of US signals, such as segmentation, morphological and predictive analysis, and research perspectives (Chap. 8).

Ultrasound Imaging: Overview and Challenges 2

We introduce medical image acquisition techniques (Sect. 2.1), ultrasound visualisation and processing (Sect. 2.2), machine learning methodologies for US image processing (Sect. 2.3), and data sets, tools, and quantitative metrics (Sect. 2.4).

2.1 Medical Imaging

The X-ray technique was the first acquisition method for biomedical imaging, developed in the late nineteenth century; the properties of the X-ray wave allow physicians to visualise the internal structures of a patient, with the subsequent introduction of contrast agents to improve soft tissue visibility. After World War II, the introduction of the ultrasound technique allowed physicians to visualise internal structures without applying ionising radiation, thus reducing the damage to human cells. US acquisition benefits many properties, such as low cost, portability, and no harm to the human body; however, ultrasound waves are less robust to perturbation than X-rays, generating noisy signals that affect the visualisation and processing of the images. In 1967, the introduction of *Computed Tomography* (CT) exploited the Radon transform to reconstruct 3D images by collecting multiple X-ray projections. The CT requires processing the acquired data for image reconstruction, thus requiring more expensive hardware and longer acquisition time than X-rays and US.

Magnetic Resonance Imaging (MRI) was introduced in the early 70s and applied strong magnetic fields to change the spin of atoms in human cells. The detection and processing of these changes are used to reconstruct 3D images of internal body structures. MRI does not apply ionising radiation, even if contrast agents can be applied to improve the visualisation of soft tissues. Generating multiple magnetic fields requires specialised hardware, making MRI an expensive acquisition method. The use of radioactively labelled markers was first

S. Cammarasana and G. Patanè, *Real-Time Processing of Ultrasound Images*, Synthesis Lectures on Visual Computing: Computer Graphics, Animation, Computational Photography and Imaging, https://doi.org/10.1007/978-3-032-05805-8_2

Table 2.1 Comparison of different biomedical acquisition techniques

Exam	Cost ($) [22]	Ionized radiation	Preparation time	Acquisition time
X-Ray	410	High	None	10 min
MRI	2048	Low	0–30 min	10 min
PET	1138	High	60 min	30–45 min
CT	1565	High	30 min	10–20 min
US	410	None	None	Real-time

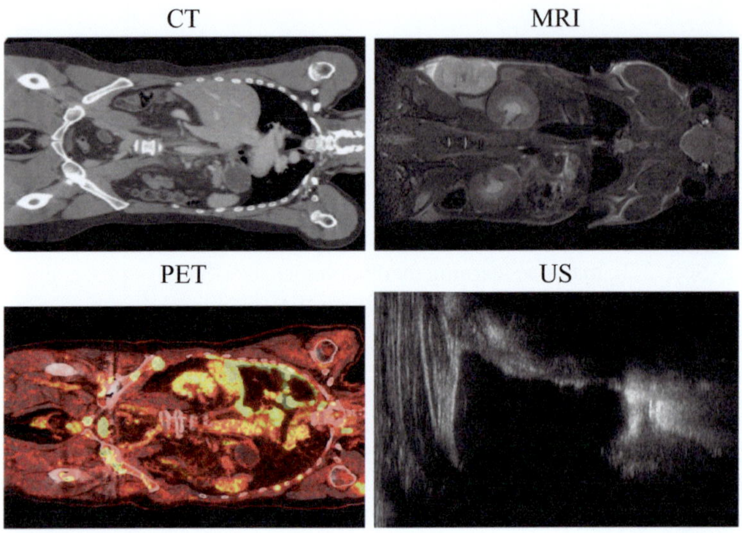

Fig. 2.1 Images acquired with different acquisition methods. Open access data set: [9, 24]

introduced in the 50s, with the significant introduction of fluorodeoxyglucose F 18 radiotracer in 1976; the *Positron Emission Tomography* (PET) accounts for the electron-positron annihilation events of the radiotracer, which becomes more robust in sites of high metabolic activity. Gathering the concentration of active radioactive compounds allows physicians to visualise physiological processes, such as the metabolic process of human cells. The joint analysis of different acquisition techniques, such as MRI-PET or CT-PET, has spread in recent years. Instead, the ultrasound examination is applied with CT or MRI through co-registration techniques to support surgery and patients' treatments. Table 2.1 summarises the main differences in cost, acquisition time, and human body damage among biomedical acquisition techniques. Figure 2.1 shows the different image outputs of the acquisition methods. We introduce the ultrasound acquisition methodology (Sect. 2.1.1), US image properties (Sect. 2.1.2), main challenges (Sect. 2.1.3), limitation (Sect. 2.1.4), and standard and security aspects (Sect. 2.1.5) of ultrasound imaging.

2.1.1 Ultrasound Acquisition

Ultrasound acquisition was applied in 1947 by Dr. Karl Theo Dussik, who built the first tool to visualise the human brain and ventricles through ultrasonic waves. Nowadays, ultrasound is widespread as an imaging technology in biomedicine (Fig. 2.2). It is cheap, portable, and free of ionising radiation compared to other imaging modalities such as CT and MR. The images are acquired in real-time, thus providing instantaneous feedback for the visual interpretation of the physician and interventional procedures such as regional anaesthesia. Modern US devices apply sound waves in the range of 1–20 MHz, and the acquisition is performed through a pulse-echo approach with a *brightness-mode* (B-mode) display: small pulses are transmitted from the ultrasound probe into the human body. The pulse is generated with two or three cycles at the same frequency, with a pulse repetition frequency (PRF) from 1 to 10 kHz (Fig. 2.3). An ultrasonic wave is a longitudinal wave with a short wavelength; it propagates in any human part and is widespread thanks to its physical properties of reflection, scattering, attenuation, and the Doppler effect. US waves have beam-emitting properties and propagate in one direction with strong directivity and in a straight line through the medium; the direction of the ultrasound wave along the beam is called the axial direction, and the direction perpendicular to the axial one is called the lateral direction.

The ultrasound waves penetrate the body structures along their path; when they pass through adjacent body parts with a different acoustic impedance, a fraction of the ultrasound pulse returns as a reflected wave, generating an echo that returns to the probe. In contrast, the rest of the wave continues to penetrate along the beam to greater tissue depths. The amplitude of the reflected echo is proportional to the difference in acoustic impedance between two

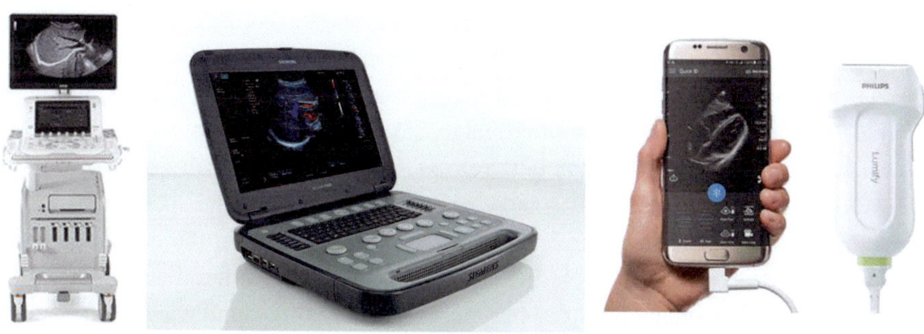

Fig. 2.2 Ultrasound acquisition machines. Left: standard (Esaote MyLab X9). Centre: portable (Siemens ACUSON P500). Right: ultrahandheld (Philips Lumify). Images courtesy of [8, 18, 21]

Fig. 2.3 Ultrasound wave emission

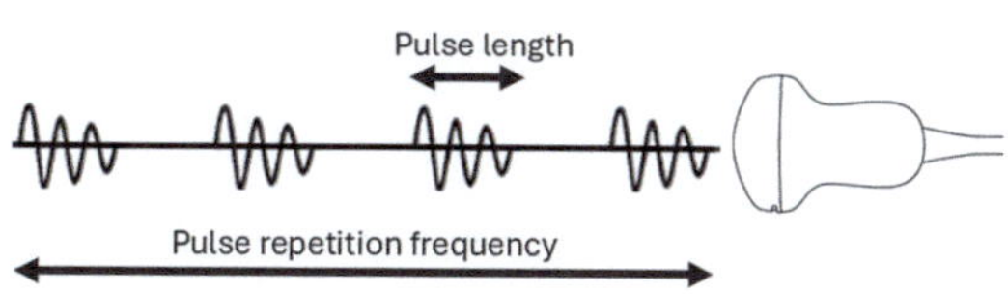

Table 2.2 Acoustic impedances of different body tissues and organs [6]

Body tissue	Acoustic impedance 10×10^6 Rayls
Air	0.0004
Lung	0.18
Fat	1.34
Kidney	1.63
Liver	1.65
Blood	1.65
Muscle	1.71
Bone	7.8

adjacent media. For example, interfaces between soft tissue and dense organs (e.g., bones) generate strong echoes due to a large acoustic impedance gradient. The acoustic impedance is a physical property of a medium defined as the density of the medium times the velocity of the wave propagation. Human body tissues have different acoustic impedances (Table 2.2): for example, air-containing organs (such as the lung) have the lowest acoustic impedance, while dense organs have a higher acoustic impedance. Ultrasound transducers (or probes) contain multiple piezoelectric crystals, which are interconnected electronically and vibrate in response to an applied electric current; the ultrasound probe works both as a speaker, generating the ultrasound wave, and a microphone, measuring the returned echo.

The reflected echoes are combined and processed to generate an image. High-frequency ultrasound waves (short wavelength) offer images of high axial resolution; in contrast, low-frequency waves (long wavelength) create images of lower resolution but can penetrate deeper structures due to a lower degree of attenuation. Defining the half-value depth as the distance at which 50% of the ultrasound energy has been dissipated, 1-MHz continuous ultrasound has a half-value depth of approximately 2.5 cm. In comparison, 3-MHz ultrasound has a half-value depth of approximately 0.8 cm at a fixed wave intensity of 1 W/cm^2 [7]. In US images, most biological tissues show scattering structures. The speckle signal that provides the visible texture in organs (e.g., liver, heart) results from the constructive and destructive interaction of the acoustic fields between multiple scattered echoes produced within the volume of the incident ultrasound pulse. Finally, ultrasound waves penetrate the human body and interact with body tissue. Ultrasound at low frequency and high intensity is used in treating biological tissues to destroy focal lesions selectively. Table 2.3 summarises ultrasound ranges and applications.

Table 2.3 Applications of US waves. Courtesy of [27]

Ultrasound waves	Classifications	Applications
Ultrasound intensity	Low-intensity ultrasound ($<3\,\mathrm{W/cm^2}$)	Therapeutic medicine, imaging medicine, medical diagnosis, and drug delivery
	High-intensity ultrasound ($\geq 3\,\mathrm{W/cm^2}$)	Surgery, cancer ablation, and palliative treatment
Ultrasound frequency	Low frequency ultrasound (20–200 kHz)	Drug delivery, surgery, cancer ablation, and palliative treatment
	Medium frequency ultrasound (0.7–3.0 MHz)	Therapeutic medicine, such as bone-fracture healing, soft-tissue lesions healing, inhibiting inflammatory responses, and erectile dysfunction treatment
	High frequency ultrasound (1–20 MHz)	Imaging medicine and medical diagnosis

2.1.2 Ultrasound Imaging

The signal is acquired by an ultrasound machine with polar coordinates, and it is transformed into an image defined on a regular grid $m \times n$, where the number of lines m corresponds to the number of sensors of the probe and the number of columns n corresponds to the depth of the acquisition. We process US signals in their Cartesian coordinates since most of the mathematical operations are straightforward in Cartesian space; after the processing, the US image is converted back into polar coordinates for a visualisation consistent with the acquisition process.

US signals can be acquired at different planes through the manual/automatic movement of the probe; furthermore, the physician can acquire both static and dynamic signals. These acquisitions are associated with various input signals: US images, dynamic 2D US images (i.e., videos), and 3D volumetric images. Each of these types of signals has its properties and advantages. US videos are acquired at different anatomical districts (e.g., abdominal, cardiac) and allow the physician to analyse the temporal variation of an anatomical feature (e.g., the movement of a muscle, the volume of the ventricle). This variation can be generated by the shift of the probe or by the movement of the anatomical part. US videos are acquired through 2D probes, which capture sequences of images at a given frequency. The resolution of each frame is affected by the required frequency of the video since some anatomical districts (e.g., cardiac) require great temporal detail to analyse better the behaviour of anatomical features that quickly change during time; in particular, the relationship between image resolution and video frequency f is given by $f = c/(2 \cdot d \cdot l)$, where c is the speed of sound, l is the number of lines, and d is the depth of the acquisition.

Volumetric data are usually acquired at obstetric anatomical district through a 2D probe, which is installed into an automatic tool; through the oscillation within a given range, equally spaced slices are acquired and stacked into the 3D volume; volumetric images are intrinsically anisotropic due to the different resolution of each slice, with respect to the resolution of the probe in its movement direction. 3D volumetric images offer a complete understanding of the spatial anatomy of the acquired region, at a more expensive tool for the acquisition. US 2D images are simple to acquire and process, but offer a limited view of the anatomical part of interest. Furthermore, the processing of US 2D images is computationally cheaper than volumetric images and videos, but cannot exploit the spatial and temporal redundancies of the signal.

Ultrasound imaging is widespread for negligible ionising radiation risks, high cost-effectiveness, portability, and rapid and real-time feedback during the evaluation of diseases in different situations, such as standard radiological examinations, pre-hospitalisation, intensive care unit, emergency room, and during surgical or interventional procedures, thus integrating the information from computed tomography or magnetic resonance. All these aspects improve the quality and accuracy of the diagnosis, reducing morbidity and mortality and monitoring healthcare costs and efficiency in primary healthcare.

The main applications of US imaging are the evaluation of inflammatory and infectious conditions, acute abdomen, cardiopulmonary function, and musculoskeletal and vascular pathologies. The main issues of the ultrasound technique are a significant loss of information during the reconstruction of the signal, the dependency of the signal on the direction of acquisition, and an underlying noise that corrupts the image and significantly affects the evaluation of the morphology of the anatomical district. In particular, speckle noise is a multiplicative noise generated when the scale of organ tissue's structures is close to the ultrasonic wavelengths; different waves are reflected and overlap, generating the perturbation in the signals appearing in the output image as darker and brighter grains. Speckle noise affects the evaluation of ultrasound images, resulting in poor image contrast degree, inadequate and non-identifiable characteristics of the tissue and structure features, and making it more challenging to identify geometries and edges of anatomical features, thus influencing the accuracy of diagnoses.

2.1.3 Challenges of US Image Processing

Several aspects affect US images, including noise, low resolution, heterogeneity, limitation in field-of-view and penetration, real-time requirements, and operator dependencies. These aspects are relevant in denoising, enhancement, and super-resolution processing.

Real-Time Ultrasound Denoising and Enhancement

Speckle noise occurs when the size of the tissue structures is dimensionally close to the wavelength of the ultrasound waves, and the reflections overlap, creating interference patterns. These noise patterns appear as dark and bright spots, affecting the image interpretation, reducing the contrast and affecting the identification of anatomical features. The *advantages of denoising* are the improved visual clarity and the enhancement of boundaries and contrast, which support the accuracy of clinical evaluations, the identification of anatomical structures and potential abnormalities, and the decision-making process of the clinicians. Noise reduction also affects post-processing techniques such as segmentation and classification, where local information is critical. The *disadvantage of denoising* is the modification of the original image and the change of the contours and appearance of anatomical structures, leading to inaccuracies in measurements and misinterpretation of clinical outcomes. Also, denoising algorithms can increase the time from data acquisition to image display, which can impact patient management in clinical settings. Several methods have been proposed for the real-time denoising of ultrasound images with a focus on different aspects of image processing, such as spatial domain filtering, non-local means filtering, spectral methods (e.g., wavelet-based), machine learning, deep learning, and low-rank methods.

A full evaluation of denoising algorithms involves both quantitative and qualitative assessment. *Quantitative metrics* such as PSNR and SSIM give the user an objective way to compare different denoising methods, while a *qualitative assessment* from ultrasound experts (e.g., clinicians, industrial specialists) is relevant to evaluate the improvement of the denoised image in terms of identification of relevant anatomical features, e.g., the boundaries of an organ. Finally, the *computational cost* and *execution time* of the denoising algorithm are relevant to match the real time industrial and clinical requirements of ultrasound acquisition.

Real-Time Ultrasound Super-Resolution

US images are characterised by lower spatial resolution with respect to other imaging modalities, such as MR and CT. The spatial resolution of ultrasound is limited by the diffraction limit that constrains the distinction between two closely spaced objects. This limit is imposed by the physics of ultrasonic wave propagation, depends on the wavelength of the ultrasound and the aperture size of the transducer, and is about half of the wavelength of the beam. Furthermore, the resolution of the acquired image is reduced when the temporal frequency of acquisition is increased. *Super-resolution* (SR) enhances the spatial resolution of the US images by reconstructing the non-acquired data. SR is significant for improving the quality of the output image, supporting diagnosis and treatment planning, and enhancing the visualisation of anatomical structures. For example, in cardiac imaging, high-resolution images allow the physician to better evaluate cardiac function, such as the variations of the ventricular volume or the movements of the mitral valve. Super-resolution is also relevant, where high temporal resolution is required to capture fast-moving structures such as blood flow. Finally, high-resolution images are relevant for the evaluation of tumours and the identification of musculoskeletal disorders.

Several methods have been proposed for the super-resolution of ultrasound images with a focus on different aspects of image processing, such as interpolation-based, learning and deep-learning-based, super-resolution accounting for multiple frames, and sparse representation and compressed sensing. The quantitative and qualitative evaluation of high-resolution images is relevant for evaluating and comparing the quality of the SR methods. Furthermore, native high-resolution images can be degraded and then reconstructed through SR methods, quantitatively comparing the ground truth with the predicted image.

2.1.4 Limitation of US Imaging

Computed tomography and magnetic resonance generate consistent and reproducible images with reduced manual intervention of the radiologic technician and, consequently, poor dependence on operator skill. In contrast, the experience and skill of the radiologist, the patient's body type, posture, and cooperation significantly affect the quality of ultrasound images. Since portable US equipment has enabled non-radiology-trained medical personnel to perform US examinations in urgent clinical situations, more quality assurance and standardised training are required. The US imaging quality is also affected by the angle of the probe touching the skin, shadowing from superficial structures, and the depth of the target lesion. These limitations and common variables, such as US operating settings and artefacts, make employing AI in US examinations challenging. Due to these complex interactions, the US image information used as input data for AI applications is subjected to random noise from unidentifiable sources. Standardised protocols for acquiring static and time-varying US images are necessary to overcome limitations in US imaging. Time-varying US images account for anatomical variations among patients and reduce the gap between those performing and interpreting the scans. However, time-varying US images often need to provide a clear depiction of the intended information due to changes in dynamic anatomy. The best results are obtained for small or fixed anatomic structures, e.g., for spatio-temporal analysis and deep learning in echocardiography and obstetric pelvic ultrasonography, joint diseases, foetal evaluation, neonatal disease, and tumour classification. In these cases, software that can overcome the limitations of US acquisition by producing standardised image data using 3D acquisition of focused anatomical areas of interest will limit noise and improve accuracy.

US acquisition has poor effectiveness on deep structures, e.g., the heart and liver, due to the attenuation of the ultrasonic waves along human tissues. Ultrasonic waves do not propagate efficiently through air, leading to limitations in the acquisition of anatomical structures such as lungs or bowel and reducing the visualisation of the underlying structures [20]. The anatomical characteristics of the patient, such as obesity or scar tissue, affect the accuracy of ultrasound acquisition. For example, the thickness of the subcutaneous fat in obese patients reduces the image quality of deeper structures [25].

2.1.5 Ultrasound Standards and Security

Nowadays, with the development of different imaging techniques (e.g., US, MR), numerical technologies are dominant in diagnostic imaging. Thus, the management of medical images is inherently based on information technologies, making computers at the core of generation, memorisation, transmission, and access to those images. In this context, it is necessary to determine a universal standard system for transmitting biomedical images. All the companies supplying equipment for the production of images apply the *Digital Imaging and Communications in Medicine* (DICOM) standard as a communication protocol to guarantee a high interconnection and interoperation capability. The DICOM standard comprises a multi-part document comprising 18 modules (Fig. 2.4). *Part 3 Information Object* definitions specify the information objects to be exchanged and the definition of the semantics of each data element. This part is very long because the various imaging modalities require many technical parameters. *Part 4 Service Class* specifications describe the services for exchanging information, like the images themselves or other helpful information to manage the images. *Part 16 Content Mapping Resource* defines the terminology, specifically how existing terminological resources can be used in DICOM and how to reuse grouped content items in DICOM Structured Reporting documents. *Part 5 Data Structure and Encoding* describes organising the information objects specified in Part 3 into a linear bitstream to be sent over a network connection or stored in a file and comprehends all aspects related to image compression. Finally, *Part 2 Conformance* specifies how to claim conformance to

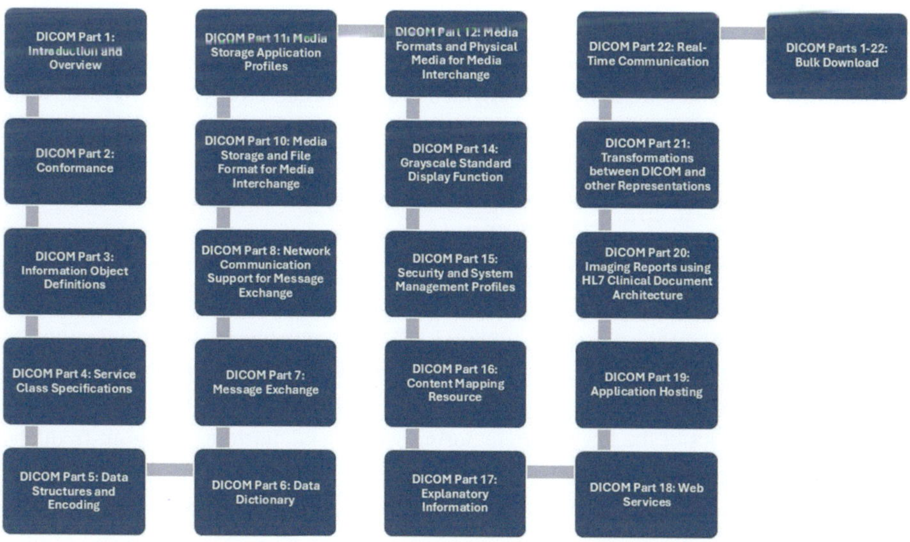

Fig. 2.4 DICOM standard multi-document organisation. Image courtesy of [11]

the DICOM standard for a particular product, meaning how to write the document called a conformance statement in detail.

This standard not only defines a transmission protocol for images but also regulates the memorisation of those images. Together with the image, this standard permits the achievement of other information identifying the patient, the exam parameters, and the image itself. Indeed, the information of the DICOM element is composed of

- a tag that identifies the element and the group to which the packet belongs. Here, the group is the DICOM entity associated with the data, and the element is the type of attribute contained in the packet. As an example, group 0100 identify the patient entity, and in this entity, element 0010 identifies the surname;
- a type indicator that identifies the format of the information memorisation;
- the byte length of the transmitted information; and
- the real principal information in the format indicated by the previous fields.

Besides the public tag, which is present in all the data saved through the standard, private tags are inserted by the specific company and contain information that depends on the particular implementation. However, such information can be fundamental for the processing and correct visualisation of the image. An in-depth analysis of DICOM standard [19] discusses several properties, such as communication protocols, storage and archiving, encoding and compression, anonymisation, encryption, and disaster-proof design.

US Data Format and Security

US signals include 2D videos of different anatomical districts (e.g., abdominal, cardiac, and obstetrical) and volumetric 3D images. 2D videos are acquired through a 2D probe, which captures sequences of images at a given frequency. The resolution of the images is affected by the acquisition frequency of the video since some anatomical districts (e.g., cardiac) require a high frequency to better analyse the behaviour of anatomical features that quickly change over time. Physicians analyse the temporal variation of anatomical features, such as the movement of a muscle or changes in the volume of a ventricle during the cardiac cycle. The analysis of the contraction and relaxation patterns of the quadriceps muscle during knee flexion and extension exercises supports the measurements of the lengthening and shortening of the muscle fibres for muscle injury diagnosis and treatment. In cardiology, specialised cardiac probes acquire 2D ultrasound videos to evaluate the function of the heart, such as the kinematics of the left ventricle during systole and diastole. The high-frequency acquisition of images allows clinicians to perform detailed visualisation of the ventricular wall movement and the opening and closing of the mitral valve to identify heart failure or valvular disorders.

Volumetric 3D images are acquired through probe oscillation within a given range; equally spaced slices are acquired and stacked into the 3D volume. Volumetric ultrasound images are significant for detailed 3D representations of anatomical structures, for example, providing a more comprehensive understanding of foetal development and maternal anatomy in the obstetric anatomical district to support the detection of anomalies and abnormalities. Furthermore, surgeons can use 3D reconstructions to evaluate the location and severity of anomalies, helping them to determine the most appropriate surgical approach or support in congenital defect repair, tumour resections, or reconstructive surgery. We encode *2D images* as $m \times n$ matrices, *2D videos* as $m \times n \times t$ tensors, where each frame has a $m \times n$ resolution, and t frames are acquired. We represent *3D (volumetric) images* as $m \times n \times p$ tensors, where p planes are acquired, and each plane has a $m \times n$ resolution. Volumetric images are intrinsically anisotropic due to the different resolutions of each slice with respect to the resolution of the probe in its movement direction.

Large biomedical data sets provide huge opportunities to improve the health sector; data analytics allow hospitals to monitor diseases and outbreaks, predict patients' behaviours, and provide risk stratification for individual patients. The processing of large amounts of data in the biomedical field must account for the privacy and protection of information. The biomedical data sets contain metadata regarding the patient's personal and clinical information, such as their identification, psychological or medical conditions, and the medical services provided. The evolution of technologies for cloud storage and online processing of data supports the development of blockchain cryptography [23] and communication protocols [1]. Some advances in biomedical privacy problems and perspectives are discussed in [15, 16].

2.2 Ultrasound Visualisation and Processing

The processing of biomedical images is an interdisciplinary field requiring medical, mathematical, informatics, physical, biological, and engineering knowledge. Applying classical computer science techniques is combined with understanding the acquisition tools and properties, the biological characteristics of the patient, and the requirements for medical post-processing. More specifically, the processing of ultrasound images faces several challenges:

- *Anatomy* of the analysed anatomical district in terms of structures, geometries and textures is relevant to evaluate the quality of the applied processing method.
- *Instrumentation* affects the output signal in terms of noise, resolution, isotropy, and artefacts; the knowledge of the physical properties of the acquisition tools allows us to optimise the application of the processing methods.
- *Acquisition modality*, such as manual versus automatic, static versus dynamic, affects the output signal regarding artefacts and processing requirements. The processing of different signals (e.g., 2D vs. 3D images) requires methods that account for the intrinsic properties of the signal.

- the *clinical evaluation* is a fundamental step for the evaluation of the performance of the processing methods due to the lack of biomedical ground truth references; for this reason, the qualitative validation of experts of US images allows us to evaluate the results in terms of preservation of the anatomical features.
- *Industrial requirements* set boundaries to the processing algorithms. Real-time calculation is the main industrial and clinical constraint in pre-processing ultrasound signals. The physician must be able to see real-time signal changes as the probe moves. In particular, the required constraint is about 30 milliseconds per image, corresponding to a frame-per-second resolution of 30 Hz. This result must be guaranteed on current ultrasonic machines equipped with modern CPUs and GPUs. We test the denoising prediction on GPU-based hardware, replicating the hardware of an ultrasound scanner.
- The *medical application* in diagnosis or intervention provides the foundation and motivation for US image analysis; the processing depends on many medical factors, such as the suspected disease, the anatomical district, and the planned post-processing/clinical steps.

2.3 Machine Learning for US Image Processing and Analysis

Large data sets of US images are available for the development of AI methods, both as traditional handcrafted feature extraction, machine learning, and deep learning techniques [13]. However, US images are complex due to their low resolution and heavily depend on the domain expert's knowledge, introducing several challenges to (i) create a well-classified, structured dataset of US images as the US image acquisition process produces much unintentional image data, and (ii) obtain a reliable training network. Deep learning is applied to US images for pattern recognition and classification based on hepatic echogenicity to diagnose fatty liver disease [9]. Furthermore, a deep-learning technique with ultrasound shear wave elastography is developed for grading hepatic fibrosis [4, 10]. AI allows clinicians to process heterogeneous data for the integration of clinical information such as family history, genetic information, menstruation status, and hormonal treatment with image analysis to improve the differentiation between benign and malignant lesions and provide predictive information for patient management. Recent advances have been made in research and clinical application of point-of-care ultrasound to solve single tasks in specific anatomic areas of interest, such as the detection of ascites, pleural effusion, pericardial effusion, pneumothorax, and pneumonia.

The application of AI methods within US images is widespread in several clinical applications for the analysis of the thyroid, breast, and prostate, as these areas are located on the surface of the body and are less affected by overlying structures and US penetration issues. AI methods are applied to accurately identify tumours and improve the diagnosis of thyroid and breast nodules, with several criticisms on variable diagnostic performance and low specificity. Quantitative colour Doppler radiomics features improve diagnostic performance for breast cancer [17]. AI is applied for scoring arthritis activity [2] and the segmentation of the cartilage [3].

2.4 US Benchmarks, Tools, Metrics

Ultrasound images are processed and compared as per their scan data format, i.e., $m \times n$ images, where the number of rows m corresponds to the number of beamlines, and the number of columns n corresponds to the depth of the acquisition (i.e., the axial resolution). In terms of visualisation, ultrasound images acquired from curvilinear probes are visualised as wedge-shaped images. For the evaluation of *quantitative metrics* between two images **N** and **p**, we account for the

- *normalised crossed correlation*

$$\text{NCC} = \frac{\sum_{i=1}^{m}[\mathbf{N}(i) - \overline{\mathbf{N}}][\mathbf{P}(i) - \overline{\mathbf{P}}]}{\left[\sum_{i=1}^{m}[\overline{\mathbf{N}} - \overline{\mathbf{N}}]^2\right]^{1/2}\left[\sum_{i=1}^{m}[\mathbf{P}_i - \overline{\mathbf{P}}]^2\right]^{1/2}},$$

  where $\overline{\mathbf{N}}$ and $\overline{\mathbf{P}}$ are the average values of the two images;
- the *mean squared error*

$$\text{MSE} = \frac{1}{m}\sum_{i=1}^{m}[\mathbf{N}(i) - \mathbf{P}(i)]^2$$

  and the *normalised mean squared error*

$$\text{NRMSE} = \left[\frac{\sum_{i=1}^{m}[\mathbf{N}(i) - \mathbf{P}(i)]^2}{\sum_{i=1}^{m}[\mathbf{N}(i)]^2}\right]^{1/2};$$

- the *peak-signal-to-noise-ration*

$$\text{PSNR} = 10\log_{10}\frac{(\max(\mathbf{N}))^2}{MSE(\mathbf{N},\mathbf{P})};$$

- the *structural similarity*

$$\text{SSIM}(\mathbf{P},\mathbf{N}) = l(\mathbf{P},\mathbf{N}) \times c(\mathbf{P},\mathbf{N}) \times s(\mathbf{P},\mathbf{N}),$$

  with $l(\mathbf{P},\mathbf{N}) = \frac{2\mu_{\mathbf{P}}\mu_{\mathbf{N}}+C_1}{\mu_{\mathbf{P}}^2+\mu_{\mathbf{N}}^2+C_1}$, $c(\mathbf{P},\mathbf{N}) = \frac{2\sigma_{\mathbf{P}}\sigma_{\mathbf{N}}+C_2}{\sigma_{\mathbf{P}}^2+\sigma_{\mathbf{N}}^2+C_2}$, and $s(\mathbf{P},\mathbf{N}) = \frac{\sigma_{\mathbf{PN}}+C_3}{\sigma_{\mathbf{P}}\sigma_{\mathbf{N}}+C_3}$, where $\mu(\cdot)$ is the mean of $(\cdot)$, $\sigma(\cdot)$ is the standard deviation of $(\cdot)$, $\sigma_{\mathbf{PN}}$ is the covariance between **P** and **N**, the positive constants C_1, C_2 and C_3 are used to avoid a null denominator;

- the *feature-based similarity* [28]

$$\text{FSIM} = \frac{\sum_{i=1}^{m} S_L(i) \cdot PC(i)}{\sum_{i=1}^{m} PC(i)},$$

where $S_L = S_{PC} \cdot S_G$ is a combination of a similarity score of the phase congruency PC [12] and the gradient magnitude G;
- the *universal image quality* (UIQ) [26]

$$\text{UIQ} = \frac{4\sigma_{\mathbf{PN}}\mu_{\mathbf{P}}\mu_{\mathbf{N}}}{(\sigma_{\mathbf{P}}^2 + \sigma_{\mathbf{N}}^2)(\mu_{\mathbf{P}}^2 + \mu_{\mathbf{N}}^2)}.$$

NCC, SSIM, and FSIM vary from 0 (worst case) to 1 (best case), PSNR varies from 0 (worst case) to $+\infty$ (best case), MSE and NRMSE go from $+\infty$ (worst case) to 0 (best case), UIQ varies from -1 (worst case) to $+1$ (best case).

The following *tools and libraries* have been used for the visualisation, processing, and enhancement of ultrasound images and videos:

- *TensorFlow* is an open-source machine learning framework developed by the Google Brain Team for building and training various machine learning models, particularly neural networks.
- *PyTorch* is an open-source machine learning library developed by Facebook's AI Research lab for the computation of image and graph features.
- *3D Slicer* is an open-source software platform for medical image informatics, image processing, and 3D visualisation, used for tasks such as image segmentation, registration, and quantitative analysis.
- *Mango* is a free, open-source medical image viewer and analysis tool for medical image data, including MRI, CT, and PET scans. Mango provides features for image viewing, such as multi-planar reformatting, intensity windowing, and basic measurements.
- *PyDICOM* is a Python package for working with DICOM files and allows the users to read, modify, and write DICOM files.
- *SimpleITK* is a simplified interface to the Insight Segmentation and Registration Toolkit (ITK) for image analysis, segmentation, registration, and filtering.
- *Field II* is a software package for simulating ultrasound imaging systems, various ultrasound transducers, tissue models, and imaging scenarios.

We account for the following *US data sets*:

- The US data set [5], which contains more than 10K ultrasound images at different resolutions and is acquired from different (e.g., musculoskeletal, obstetric, abdominal) anatomical districts. Furthermore, it includes 2D videos from different anatomical districts (e.g., cardiac, abdominal) and 3D images from the obstetric district.

- The trackless 3D freehand ultrasound reconstruction (TUS-REC) data set [14] is an open data set that provides in vivo US data acquired from both left and right forearms of one hundred volunteers (1200 scans and 1,206,900 frames), with different numbers of slices per volume, acquisition (i.e., parallel or perpendicular) and direction (e.g., distal to proximal); the large variety of data allows us to test the generality our SR method. The acquisition is tracked by a time-synchronised optical tracker for the reconstruction of the 2D US images into a 3D volume, which supports quantitative biometric measurement, 3D visualisation, and interventional guidance.

References

1. H.A. Al Hamid et al., A security model for preserving the privacy of medical big data in a healthcare cloud using a fog computing facility with pairing-based cryptography. IEEE Access **5**, 22313–22328 (2017)
2. J.K.H. Andersen et al., Neural networks for automatic scoring of arthritis disease activity on ultrasound images. RMD Open **5**(1), e000891 (2019)
3. M. Antico et al., Deep learning-based femoral cartilage automatic segmentation in ultrasound imaging for guidance in robotic knee arthroscopy. Ultrasound Med. & Biol. **46**(2), 422–435 (2020)
4. L.J Brattain et al., Image processing pipeline for liver fibrosis classification using ultrasound shear wave elastography. Ultrasound Med. & Biol. **46**(10), 2667–2676 (2020)
5. S. Cammarasana, P. Nicolardi, G. Patanè, Real-time denoising of ultrasound images based on deep learning. Med. & Biol. Eng. & Comput. 1–16 (2022)
6. V.W.S. Chan, S. Abbas, Ultrasound Imaging for Regional Anesthesia: a Practical Guide. Ultrasound for Regional Anesthesia (2009)
7. D.O. Draper, J.C. Castel, D. Castel, Rate of temperature increase in human muscle during 1 MHz and 3 MHz continuous ultrasound. J. Orthop. & Sports Phys. Therapy **22**(4), 142–150 (1995)
8. Esaote Mylab-9, https://www.esaote.com/ultrasound/ultrasoundsystems/p/mylab-9/ Accessed 17 Dec 2024
9. S. Gatidis et al., A whole-body FDG-PET/CT dataset with manually annotated tumor lesions. Sci. Data **9**(1), 601 (2022)
10. I. Gatos et al., A machine-learning algorithm toward color analysis for chronic liver disease classification, employing ultrasound shear wave elastography. Ultrasound Med. & Biol. **43**(9), 1797–1810 (2017)
11. B. Gibaud, The DICOM standard: a brief overview, in *Molecular Imaging: Computer Reconstruction and Practice* (2008), pp. 229–238
12. P. Kovesi, Image features from phase congruency. Videre: J. Comput. Vis. Res. **1**(3), 1–26 (1999)
13. Y. LeCun, Y. Bengio, G. Hinton, Deep learning. Nature **521**(7553), 436–444 (2015)
14. Q. Li et al., Long-term dependency for 3D reconstruction of freehand ultrasound without external tracker. IEEE Trans. Biomed. Eng. **71**(3), 1033–1042 (2024)
15. B.A. Malin, K.E. Emam, C.M. O'Keefe, Biomedical data privacy: problems, perspectives, and recent advances. J. Amer. Med. Inf. Ass. **20**(1), 2–6 (2013)
16. B. Mittelstadt, The ethics of biomedical 'big data' analytics. Philosop. & Technol. **32**(1), 17–21 (2019)

17. A.F. Moustafa et al., Color doppler ultrasound improves machine learning diagnosis of breast cancer. Diagnostics **10**(9), 631 (2020)
18. P. Lumify, https://www.philips.it/healthcare/sites/lumify. Accessed 17 Dec 2024
19. O.S. Pianykh, *DICOM Security* (Springer, 2008)
20. C.M. Rumack, D. Levine, *Diagnostic Ultrasound* (Elsevier Health Sciences, 2023)
21. Siemens ACUSON P500, https://www.siemens-healthineers.com/it/ultrasound/ultrasound-point-of-care/acuson-p500-ultrasound-machine. Accessed 17 Dec 2024
22. C.L. Sistrom, N.L. McKay, Costs, charges, and revenues for hospital diagnostic imaging procedures: differences by modality and hos- pital characteristics. J. Am. Coll. Radiol. **2**(6), 511–519 (2005)
23. M. Soni, D.K. Singh, Blockchain-based security & privacy for biomedical and healthcare information exchange systems, in *Materials Today: Proceedings* (2021)
24. S. Stauffer et al., Preclinical therapeutic efficacy of RAF/MEK/ERK and IGF1R/AKT/mTOR inhibition in Neuroblastoma. Cancers **16**(13), 2320 (2024)
25. H.S. Thomsen, J.A.W. Webb, et al., *Contrast Media: Safety Issues and ESUR Guidelines* (Springer, 2009)
26. Z. Wang, A.C. Bovik, A universal image quality index. Signal Proc. Lett. **9**(3), 81–84 (2002)
27. Z. Xin et al., Clinical applications of low-intensity pulsed ultrasound and its potential role in urology. Transl. Androl. Urology **5**(2), 255 (2016)
28. L. Zhang et al., FSIM: a feature similarity index for image quality assessment. Trans. Image Proc. **20**(8), 2378–2386 (2011)

US Denoising 3

For the US visualisation based on denoising and enhancement, it is relevant to address speckle noise, which is a multiplicative noise generated when the scale of organ tissue's structures is close to the ultrasonic wavelengths; different waves are reflected and overlap, causing the perturbation in the signals and appearing in the output image as darker and brighter grains. Ultrasonic image denoising has pros and cons: noise reduction improves visual evaluation, and the application of post-processing methods that use local information and are therefore influenced by noisy elements that alter the accuracy of numerical operators, such as convolution and interpolation. We discuss the main properties of ultrasound image denoising (Sect. 3.1). Then, we review state-of-the-art denoising methods classified according to their class: non-local (Sect. 3.2), anisotropic (Sect. 3.3), spectral (Sect. 3.4), learning-based (Sect. 3.5), and low-rank (Sect. 3.6). Finally, we present a comparison of the presented methods in terms of denoising results and execution time (Sect. 3.7).

3.1 US Denoising: Main Properties

Denoising is relevant and widespread in US processing, and many works propose various appropriate denoising methods to address the specific features of ultrasonic images. Ultrasonic image processing through denoising has pros and cons: noise reduction improves visual evaluation, and the application of post-processing methods that use local information and are therefore influenced by noisy elements that alter the accuracy of numerical operators (e.g., convolution) and methods (e.g., interpolation). At the same time, applying a denoising algorithm means changing the image and, therefore, potentially, the contours and anatomical structures. Additionally, image denoising increases the time from data acquisition to screen display for the physician. Possibly, joint analysis of the two images (i.e., noisy and

S. Cammarasana and G. Patanè, *Real-Time Processing of Ultrasound Images*, Synthesis Lectures on Visual Computing: Computer Graphics, Animation, Computational Photography and Imaging, https://doi.org/10.1007/978-3-032-05805-8_3

denoised) allows the physician to have a complete picture and improve the diagnosis. For these reasons, a valid denoising algorithm must satisfy the following properties:

- *smoothing* of pixels with high-gradient features and *signal-driven* denoising, taking into account the spatiotemporal properties of the input data, including 2D and 3D images and 2D videos;
- *reduction* of artefacts and blurring effects and *preservation* of anatomical structures, edges, geometries, and texture properties;
- *computational cost* compatible with real-time visualisation constraint and *memory consumption* compatible with hardware requirement and *hardware* requirements, regarding CPUs and GPUs computing power;
- *versatility* with different anatomical districts and *adaptability* to the visual perception of a different physician through the proposal of soft and hard smoothing; and
- *qualitative and clinical validation* by an industrial or medical expert due to the lack of ground truth references for evaluating the denoising properties.

Denoising ultrasound images is relevant for post-processing and visual evaluation by medical experts. Despite some relevant works consider raw ultrasound images and videos for cardiac segmentation [30, 39], several results show the benefits of denoising for *segmentation* [53, 60, 65], *feature extraction* [23], *classification* [49, 56], *super-resolution* [25], *registration* [14], and *texture analysis* [47]. Furthermore, main ultrasound machine manufacturers include a denoising filter in their scanners [18, 45]. Regarding hardware requirements, several ultrasound machines manufactured by main competitors (e.g., Esaote, Philips) are equipped with GPU cards [18, 44]. Furthermore, some recent denoising methods for ultrasound images have been developed on GPUs [5, 40]. Finally, the application of GPUs to image processing for future medical ultrasound imaging systems [50] presents the advantages of GPUs over CPUs in terms of performance, power consumption, and cost.

We introduce non-local (Sect. 3.2), anisotropic (Sect. 3.3), spectral (Sect. 3.4), learning-based (Sect. 3.5), and low-rank (Sect. 3.6) denoising methods. Then, we compare denoising results (Sect. 3.7) in terms of quantitative metrics and qualitative evaluation.

3.2 Non-local Denoising

The *Non-Local Means* (NLM) denoising [7] uses the patterns' redundancy of the input image, and each patch is restored with a weighted average of all the other patches, where each weight is proportional to the similarity among the patches. The *Bayesian non-local mean filter* [24] improves the NLM with the introduction of a Bayesian estimator as a distance measure among the patches, which allows the user to better determine the amount of denoising by the noise variance of the patch. The anisotropic neighbourhood in NLM [36] uses an image gradient to estimate the edge orientation and then adapts the patches to match the local edges. The characterisation of the patches through a redundancy index [37] improves the self-similarity computation among patches. The improvement in the structure of the search window is achieved through the computation of an optimal search window for each pixel [54], according to the denoising degree of the related patch.

3.3 Anisotropic Denoising

The denoised image is computed as the solution to an anisotropic diffusion equation [42, 43], where the gradient of the image guides the diffusion process. The variant [62] exploits the Lee [28] and Forst [20] filters, which are edge-sensitive to speckle noise. An improvement of the previous results [2] is achieved by applying the Kuan filter [27] in the diffusion equation and revising the neighbourhood selection used for the statistical parameter estimation. The anisotropic denoising introduces a class of fractional-order anisotropic diffusion equations [3], using the Fourier transform to compute the fractional derivatives and the discrete Fourier transform to calculate the fractional-order differences.

3.4 Spectral Denoising

Denoising based on *spectral decomposition* transforms a signal into its spectral domain and exploits the sparsity of the converted signal to remove noise through a threshold operation. Several transformations have been applied to image denoising, such as Wavelets [10, 31, 38, 46], Curvelets [52], Contourlets [11], and Shearlets [61]. To reconstruct the denoised image, the *3D block-matching* [13] computes and stacks similar patches through NLM; each stack is transformed into its spectral domain with wavelet decomposition, denoised through a hard/soft threshold, and reconstructed in the space domain. A collaborative filter aggregates the denoised patches. The *synthetic aperture radar block matching 3D* (SAR-BM3D) [41] introduces a speckle-based variant of 3D block matching; the similarity among the patches is computed by considering the probability distribution of the speckle noise as a distance metric. Furthermore, the hard/soft threshold of the wavelet transformed signal is substituted by a Local Linear Minimum Mean Square Error (LLMMSE) filter. The *principal component analysis block matching 3D* (PCA-BM3D) [12] improves the stacking operation of 3D block-matching by using shape-adaptive neighbourhoods, which enables its local adaptability to image features. The 3D transformation of each stack to the spectral domain is performed through the PCA [57] and an orthogonal 1D transformation in the third dimension. Finally, the *industrial algorithm* is a private method that accounts for the recursive pyramid algorithm for the discrete wavelet transform [55].

3.5 Learning-Based Denoising

A *learned simultaneous sparse coding* method [35] integrates sparse dictionary learning with non-local self-similarities of natural images. The *non-locally centralised sparse representation* (NCSR) [16] exploits the non-local redundancies, combined with local sparsity properties, to estimate the coefficients of the sparse representation of the input image. The dictionary is learned by clustering the patches of the image into K clusters through the K-means [34] method and then learning a PCA sub-dictionary for each cluster. This method has been further improved in [59] with a fast version based on a pre-learned dictionary and improving computational efficiency. The structured sparse model selection over a family of learned orthogonal bases [33] is applied to deblurring images with Gaussian noise. In the *Noise2Noise* algorithm [29], the network learns to denoise images only considering the noisy data without knowing the ground truth. The *Noise2Void* algorithm [26] further expands this idea and does not require a couple of noisy images for the training. This approach is relevant in biomedical fields, where no ground truth images exist. The *Noise2Self* method [4] proposes a self-supervised algorithm that does not require any prior information on the input image, estimation of the noise, or ground truth data. The denoising of images [19] is achieved by extracting features from the noisy image through a *Convolutional Neural Network* (CNN) and combining the edge regularisation with the total variation regularisation. The combination of CNN and low-rank representation [21] is applied to detect anomalous pixels in hyperspectral images. The multilevel wavelet convolutional neural network [58] is used for restoring blurred images affected by Cauchy noise. The *block matching Convolutional Neural Network* (BM-CNN) [1] integrates a deep learning approach with the 3D block-matching method; the denoising of the stacks is predicted through a DnCNN [63], which is trained with a data set of 400 images corresponding to more than 250K training samples. A feed-forward CNN smooths images independently from the noise level by exploiting residual learning and batch normalisation. Then, the blocks are aggregated, and the image is reconstructed, as in the 3D block-matching algorithm.

3.6 Low-Rank Denoising

Low-rank approximation computes the denoised image as the solution to a weighted minimisation problem, whose cost function is the Frobenius norm [8, 42, 51] or the ℓ_1 norm [17], between the input and the target images. The relation between local and non-local information [15] allows us to estimate signal variances by interpreting the *Singular Value Decomposition* (SVD) through a bilateral variance estimation. In [48], a high-order SVD is applied to 3D blocks, and the denoised image is achieved with hard thresholding of the decomposed signal. The *Weighted Nuclear Norm Minimisation* (WNNM) [22] computes the stacks as in the 3D block-matching method, performs the SVD on the stacks, and applies a weighted threshold to the singular values, where higher weights correspond to lower singular values, which capture the noisy component of the image. The *collaborative filtering* of WNNM for aggregating the denoised patches is performed as in the 3D block-matching method. The weighted nuclear norm and the histogram preservation [64] are combined in a single constrained optimisation problem, which is solved through the alternating direction method of multipliers [6]. The WNNM is extended to image deblurring with several types of noise [32].

To improve the quality of the denoised image, the tuned-WNNM optimises the WNNM parameters and is more suited for US images. Given a pixel x, the *patch* P_x is the set of pixels in the neighbourhood of x; each pixel of the image has a related patch. The *search window* is the set of patches selected for searching the closest ones to a reference patch under a certain metric. The *stack*, or *3D block*, is the set of patches similar to a reference patch; these patches are stored in a 3D structure, and the redundancy of the stack is exploited to remove the noise. Within this setting, the following parameters are identified and optimised: the number of patches is no longer limited by the step value (e.g., 1 patch every 2 or 3 pixels), and a patch for each pixel is assigned; this parameter allows the user to increase the number of processed patches, thus improving the data redundancy. The block-matching algorithm is performed at every iteration instead of once every two iterations; the selection of the searching window and the size of the stack is larger than in *WNNM*. These parameters allow the user to improve the similarity measure among 3D blocks and the accuracy of the denoising method.

The tuned-WNNM method is also specialised for ultrasound images by varying the denoising intensity through a parameter that affects the threshold of the singular values of the SVD. Implementing a computationally optimised and potentially real-time version of the tuned WNNM is a very tough requirement; the main iterative cycle of the algorithm is not parallelisable due to the dependency of the data among the iterations. Furthermore, the cubic computational cost for evaluating the SVD of each block is no longer optimisable. The real-time requirement needs a specific hardware-based implementation, and any modification to the method requires a new parallel algorithm implementation. This approach needs dedicated and more expensive hardware, which contrasts with the cheapness of ultrasound acquisition.

3.7 Experimental Results

We evaluate the denoising results on different anatomical districts of the US data set (Figs. 3.1, 3.2, 3.3 and 3.4): WNNM, NCSR, and PCA-BM3D have been judged as the best methods in terms of denoising, and WNNM outperforms all the other methods in terms of edge preservation and enhancement. In particular, WNNM well preserves the edges of the muscular fibres (Figs. 3.1 and 3.4) and the internal organs (Figs. 3.2 and 3.3). The output of SAR-BM3D shows a granular effect, which negatively affects the preservation of the anatomical features, and BM-CNN generates artefacts, which are typical of a deep learning approach. According to these results, we selected WNNM as the best denoising ultrasound image. However, we underline that the other methods have denoising and edge preser-

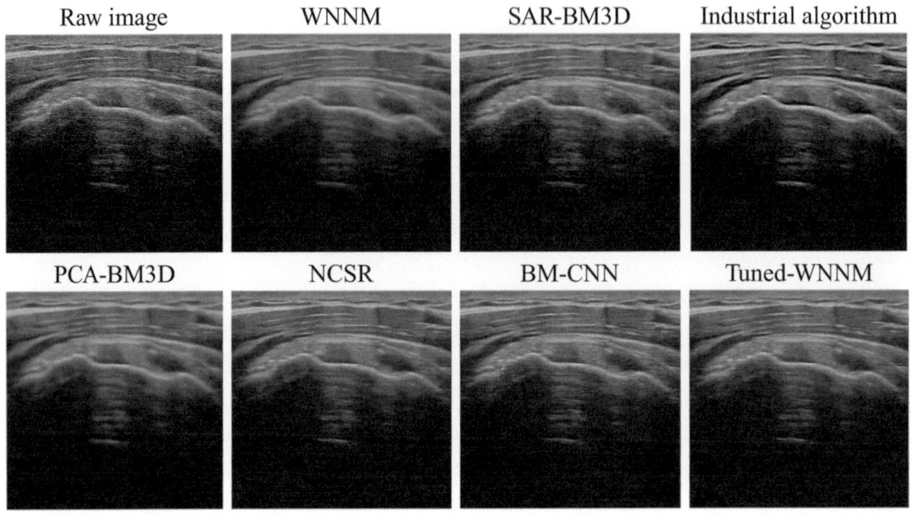

Fig. 3.1 Raw images of a musculoskeletal district and denoised images, visualised in a scan-converted format. Image courtesy of [9]

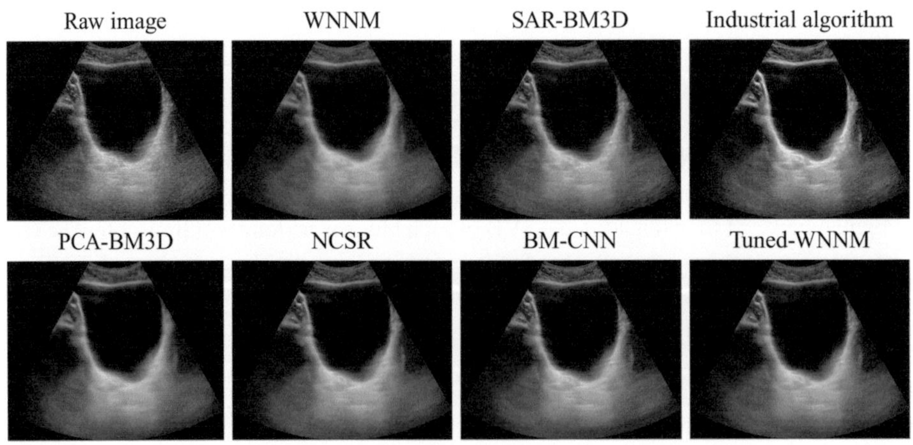

Fig. 3.2 Raw data set of an abdominal district and denoised images, visualised in a scan-converted format. Image courtesy of [9]

vation characteristics, which could be included in the framework as alternative denoising algorithms.

Then, we compare the denoising results of *tuned-WNNM* (Figs. 3.1, 3.2, 3.3 and 3.4), bottom-right. Also, we analyse the role of the parameter that affects the threshold of the singular values of the SVD and, consequently, the intensity of the smoothing. Increasing this parameter, the method improves in terms of removed noise, though introducing a low

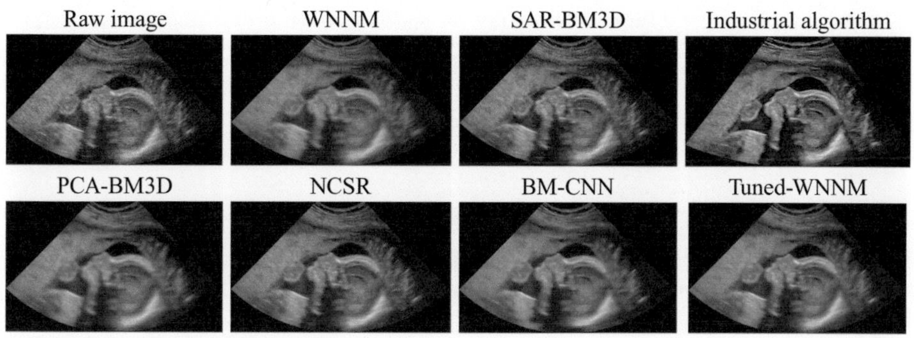

Fig. 3.3 Raw data set of an obstetric district and denoised images, visualised in a scan-converted format. Image courtesy of [9]

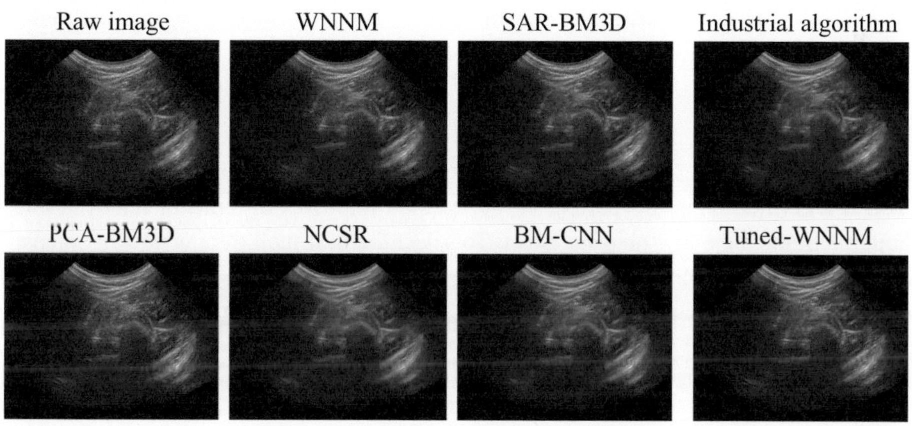

Fig. 3.4 Raw data set of arm district and denoised images, visualised in a scan-converted format. Image courtesy of [9]

blurring effect. To select the best tuning for denoising intensity, we choose the output image that best fits the medical requirements among three different levels of denoising intensity (Fig. 3.5). In particular, Fig. 3.5b shows the best result as a compromise between noise removal, edge preservation, and blurring effect; in fact, it preserves the geometry of the internal tissues while enhancing the edges of the anatomical structures.

Execution Time

Our tests (Table 3.1) are executed with MATLAB R2020a on a workstation with 2 Intel i9-9900KF CPUs (3.60 GHz), 32 GB RAM, and none of these methods achieve real-time computation. In particular, WNNM takes more than three minutes to process a 600×485 image, and the fastest method (i.e., SAR-BM3D) takes about one minute; however, real-time computation in an ultrasound environment requires a processing time in the order of a few milliseconds. This result motivates the development of a deep learning framework for

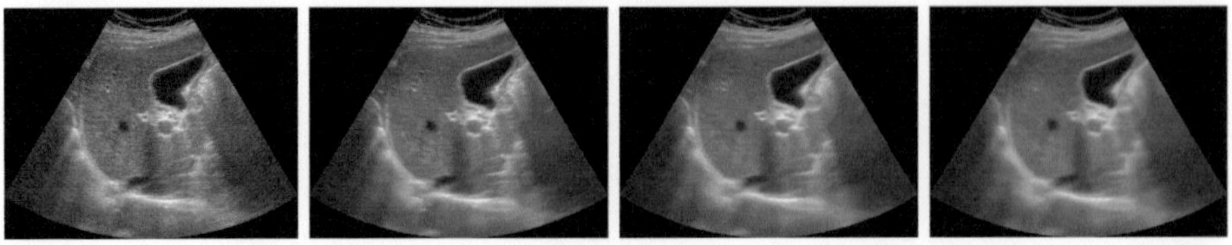

Fig. 3.5 Ultrasound image of an abdominal district and denoised images achieved by applying the tuned-WNNM and varying the denoising intensity from low (**a**) to high (**c**). Image courtesy of [9]

Table 3.1 Execution time computed as an average value of a set of US images at 600 × 485 resolution

Method	WNNM	SAR-BM3D	BM-CNN	NCSR	PCA-BM3D	Industrial algorithm
Execution time [s]	215	**55**	356	380	95	Real-time

the real-time denoising of ultrasound images, further optimised with an HPC framework (Sect. 5.1.2). Increasing this parameter, the method improves in terms of removed noise, though introducing a low blurring effect. To select the best tuning for denoising intensity, we choose the output image that best fits the medical requirements among three different levels of denoising intensity (Fig. 3.5). In particular, Fig. 3.5b shows the best result as a compromise between noise removal, edge preservation, and blurring effect; in fact, it preserves the geometry of the internal tissues while enhancing the edges of the anatomical structures.

References

1. B. Ahn, N.I. Cho, Block-matching convolutional neural network for image denoising (2017). arXiv:1704.00524
2. S. Aja-Fernandez, C.A.-L. Opez, On the estimation of the coefficient of variation for anisotropic diffusion speckle filtering. IEEE Trans. Image Process. **15**(9), 2694–2701 (2006)
3. J. Bai, X.-C. Feng, Fractional-order anisotropic diffusion for image denoising. IEEE Trans. Image Process. **16**(10), 2492–2502 (2007)
4. J. Batson, L. Royer, Noise2Self: blind denoising by self-supervision (2019). arXiv:1901.11365
5. B. Biswas, B.K. Sen, K.N. Dey, Ultrasound medical image deblurring and denoising method using variational model on CUDA. Advanced Computing and Systems for Security. Springer (2018), pp. 95–108
6. S. Boyd et al., Distributed optimization and statistical learning via the alternating direction method of multipliers. Found. Trends Mach. Learn. **3**(1), 1–122 (2011). ISSN: 1935-8237
7. A. Buades, B. Coll, J.-M. Morel, A non-local algorithm for image denoising. In: IEEE Computer Society Conference on Computer Vision and Pattern Recognition, vol. 2, pp. 60–65 (2005). https://doi.org/10.1109/CVPR.2005.38

8. S. Cammarasana, P. Nicolardi, G. Patane, A universal deep learning framework for real-time denoising of ultrasound images (2021). arXiv:2101.09122
9. S. Cammarasana, P. Nicolardi, G. Patane, Real-time denoising of ultrasound images based on deep learning. Med. & Biolog. Eng. & Comput. 1–16 (2022)
10. S.G. Chang, B. Yu, M. Vetterli, Adaptive wavelet thresholding for image denoising and compression. IEEE Trans. Image Proc. **9**(9), 1532–1546 (2000)
11. A.L. Da Cunha, J. Zhou, M.N. Do, The nonsubsampled contourlet transform: theory, design, and applications. IEEE Trans. Image Proc. **15**(10), 3089–3101 (2006)
12. K. Dabov et al., BM3D image denoising with shape-adaptive principal component analysis, in *Signal Processing with Adaptive Sparse Structured Representations*, ed. by R. Gribonval (2009)
13. K. Dabov et al., Image denoising with block-matching and 3D filtering, in *Image Processing: Algorithms and Systems, Neural Networks, and Machine Learning*, vol. 6064 (SPIE, 2006), p. 606414
14. T. De Silva et al., 2D–3D rigid registration to compensate for prostate motion during 3D TRUS-guided biopsy. Med. Phys. **40**(2), 022904 (2013)
15. W. Dong, G. Shi, X. Li, Nonlocal image restoration with bilateral variance estimation: a low-rank approach. IEEE Trans. Image Process. **22**(2), 700–711 (2012)
16. W. Dong et al., Nonlocally centralized sparse representation for image restoration. IEEE Trans. Image Process. **22**(4), 1620–1630 (2012)
17. A. Eriksson, A. Van Den Hengel, Efficient computation of robust low-rank matrix approximations in the presence of missing data using the L1 norm, in *IEEE Conference on Computer Vision and Pattern Recognition* (2010), pp. 771–778
18. Esaote Mylab-9, https://www.esaote.com/ultrasound/ultrasoundsystems/p/mylab 9/. Accessed 17 Dec 2024
19. Y. Fang, T. Zeng, Learning deep edge prior for image denoising. Comput. Vis. Image Underst. **200**, 103044 (2020). ISSN: 1077-3142
20. V.S. Frost et al., A model for radar images and its application to adaptive digital filtering of multiplicative noise. IEEE Trans. Pattern Anal. Mach. Intell. **4**(2), 157–166 (1982)
21. F. Xiyou et al., Hyperspectral anomaly detection via deep plug-and-play denoising CNN regularization. IEEE Trans. Geosci. Remote Sens. **59**(11), 9553–9568 (2021)
22. S. Gu et al., Weighted nuclear norm minimization with application to image denoising, in *Proceedings of the IEEE Conference on Computer Vision and Pattern Recognition* (2014), pp. 2862–2869
23. D.K. Iakovidis, E.G. Keramidas, D. Maroulis, Fuzzy local binary patterns for ultrasound texture characterization, in *International Conference Image Analysis and Recognition* (Springer, 2008), pp. 750–759
24. C. Kervrann, J. Boulanger, Pierrick Coupé, Bayesian non-local means filter, image redundancy and adaptive dictionaries for noise removal, in *International Conference on Scale Space and Variational Methods in Computer Vision* (Springer, 2007), pp. 520–532
25. P. Khavari, A. Asif, H. Rivaz, Non-local super resolution in ultrasound imaging, in *20th International Workshop on Multimedia Signal Processing* (IEEE, 2018), pp. 1–6
26. A. Krull, T.-O. Buchholz, F. Jug, Noise2Voidlearning denoising from single noisy images, in *Proceedings of the IEEE Conference on Computer Vision and Pattern Recognition* (2019), pp. 2129–2137
27. D.T. Kuan et al., Adaptive noise smoothing filter for images with signaldependent noise. IEEE Trans. Pattern Anal. Mach. Intell. **7**(2), 165–177 (1985)
28. J. Lee, Digital image enhancement and noise filtering by use of local statistics. IEEE Trans. Pattern Anal. Mach. Intell. **2**(2), 165–168 (1980)
29. J. Lehtinen et al., Noise2Noise: learning image restoration without clean data (2018). arXiv:1803.04189

30. X. Liu et al., Deep learning-based automated left ventricular ejection fraction assessment using 2-D echocardiography. Amer. J. Physiol.-Heart Circul. Physiol. **321**(2), H390–H399 (2021)
31. X. Liu et al., Efficient single image dehazing and denoising: an efficient multi-scale correlated wavelet approach. Comput. Vis. Image Underst. **162**, 23–33 (2017)
32. L. Ma, X. Li, T. Zeng, Low rank prior and total variation regularization for image deblurring. J. Sci. Comput. **70**(3), 1336–1357 (2017)
33. L. Ma, T. Zeng, Image deblurring via total variation based structured sparse model selection. J. Sci. Comput. **67**(1), 1–19 (2016)
34. J. MacQueen, Some methods for classification and analysis of multivariate observations. Proc. Symp. Math. Stat. Probab. **1**(14), 281–297 (1967)
35. J. Mairal et al., Non-local sparse models for image restoration, in *IEEE International Conference on Computer Vision* (2009), pp. 2272–2279
36. A. Maleki, M. Narayan, R.G. Baraniuk, Anisotropic nonlocal means denoising. Appl. Comput. Harm. Anal. **35**(3), 452–482 (2013)
37. F. Mei, D. Zhang, Y. Yang, Improved non-local self-similarity measures for effective speckle noise reduction in ultrasound images. Comput. Methods Progr. Biomed. **196**, 105670 (2020). ISSN: 0169-2607
38. M.K. Mihcak et al., Low-complexity image denoising based on statistical modeling of wavelet coefficients. Signal Proc. Lett. **6**(12), 300–303 (1999)
39. D. Ouyang et al., Video-based AI for beat-to-beat assessment of cardiac function. Nature **580**(7802), 252–256 (2020)
40. F.P. Xavier, de Fontes, et al., Real time ultrasound image denoising. J. Real-Time Image Proc. **6**(1), 15–22 (2011)
41. S. Parrilli et al., A nonlocal SAR image denoising algorithm based on LLMMSE wavelet shrinkage. IEEE Trans. Geosci. Remote Sens. **50**(2), 606–616 (2011)
42. G. Patane, Diffusive smoothing of 3D segmented medical data. J. Adv. Res. **6**(3), 425–431 (2015)
43. P. Perona, J. Malik, Scale-space and edge detection using anisotropic diffusion. IEEE Trans. Pattern Anal. Mach. Intell. **12**(7), 629–639 (1990)
44. Philips EPIQ Elite, Accessed 11 Feb 2022
45. Philips XRES Pro. https://www.usa.philips.com/healthcare/education-resources/technologies/ultrasound/xres. Accessed: 11 Feb 2022
46. J. Portilla et al., Image denoising using scale mixtures of Gaussians in the wavelet domain. IEEE Trans. Image Process. **12**(11), 1338–1351 (2003)
47. M. Puri et al., Texture analysis of foot sole soft tissue images in diabetic neuropathy using wavelet transform. Med. & Biolog. Eng. & Comput. **43**(6), 756–763 (2005)
48. A. Rajwade, A. Rangarajan, A. Banerjee, Image denoising using the higher order singular value decomposition. IEEE Trans. Pattern Anal. Mach. Intell. **35**(4), 849–862 (2012)
49. R. Sivanandan, J. Jayakumari, A new CNN architecture for efficient classification of ultrasound breast tumor images with activation map clustering based prediction validation. Med. & Biolog. Eng. & Comput. **59**(4), 957–968 (2021)
50. H. So et al., Medical ultrasound imaging: to GPU or not to GPU? Micro **31**(5), 54–65 (2011)
51. N. Srebro, T. Jaakkola, Weighted low-rank approximations, in *Proceedings of the 20th International Conference on Machine Learning* (2003), pp. 720–727
52. J.-L. Starck, E.J. Candes, D.L. Donoho, The curvelet transform for image denoising. IEEE Trans. Image Proc. **11**(6), 670–684 (2002)
53. J. Tang et al., Speckle reducing bilateral filter for cattle follicle segmentation. BMC Genomics **11**(2), 1–9 (2010)
54. R. Verma, R. Pandey, Adaptive selection of search region for NLM based image denoising. Optik **147**, 151–162 (2017)

55. M. Vishwanath, The recursive pyramid algorithm for the discrete wavelet transform. Trans. Signal Proc. **42**(3), 673–676 (1994)
56. M. Wei et al., A benign and malignant breast tumor classification method via efficiently combining texture and morphological features on ultrasound images. Comput. Math. Methods Med. **2020** (2020)
57. S. Wold, K. Esbensen, P. Geladi, Principal component analysis. Chemom. Intell. Lab. Syst. **2**(1–3), 37–52 (1987)
58. W. Tingting et al., Deep multi-level wavelet-CNN denoiser prior for restoring blurred image with Cauchy noise. Signal Proc. Lett. **27**, 1635–1639 (2020)
59. X. Shaoping, X. Yang, S. Jiang, A fast nonlocally centralized sparse representation algorithm for image denoising. Signal Process. **131**, 99–112 (2017)
60. F. Yang et al., A shape-optimized framework for kidney segmentation in ultrasound images using NLTV denoising and DRLSE. Biomed. Eng. Online **11**(1), 1–13 (2012)
61. H.-Y. Yang et al., Image denoising using nonsubsampled shearlet transform and twin support vector machines. Neural Netw. **57**, 152–165 (2014)
62. Y. Yu, S.T. Acton, Speckle reducing anisotropic diffusion. IEEE Trans. Image Proc. **11**(11), 1260–1270 (2002)
63. K. Zhang et al., Beyond a Gaussian denoiser: residual learning of deep CNN for image denoising. IEEE Trans. Image Process. **26**(7), 3142–3155 (2017)
64. M. Zhang, C. Desrosiers, Structure preserving image denoising based on low-rank reconstruction and gradient histograms. Comput. Vis. Image Underst. **171**, 48–60 (2018)
65. Z. Zhuang et al., Application of fractal theory and fuzzy enhancement in ultrasound image segmentation. Med. & Biol. Eng. & Comput. **57**(3), 623–632 (2019)

4 Learning-Based Low-Rank Denoising

We discuss a *learning-based low-rank* (LLR) denoising method with an application to ultrasound images. Given a training data set of ground truth images (Fig. 4.1), we apply different types of noise and intensities (e.g., Gaussian, salt and pepper, Poisson, exponential, speckle) and extract 3D blocks through the block matching algorithm [5]. Then, LLR computes the optimal singular values through a proper optimisation applied to the SVD of each 3D block. LLR iterates this approach, where the input image of each iteration is the denoised image at the previous step. The input and optimal singular values compose the training data set of the learning model; the singular values of the training data set are classified according to four parameters: noise type and intensity, number of iterations of the method, and clustering of the singular values. This classification allows the user to design specific networks and improve the accuracy of the learning model. The learning phase optimises a matrix of weights, is applied to train the model and predicts the optimal thresholds of the SVD. For image denoising, LLR computes the 3D blocks and the SVD, the four parameters of each 3D block, and the thresholds through the network's prediction. Finally, the block-matching aggregation is applied to reconstruct the denoised image.

LLR trains the learning-based model to predict the optimal thresholds of the SVD. In contrast, learning-based methods [20] predict the denoised image; low-rank methods [13] manually select the SVD thresholds; the *block-matching 3D* (BM3D) [5] applies a wavelet decomposition, with a manual selection of the thresholds. The denoising is independent of the type of noise and provides an upper bound to the accuracy of the denoising of 2D images through the SVD. This approach can be generalised to different transformations of the noisy signal, e.g., spectral (wavelet, shearlet), factorisation (e.g., SVD, HOSVD), or to volumetric data through spherical harmonics and Laplacian eigenpairs. Trained networks are available at https://github.com/cammarasana123/denoise.

S. Cammarasana and G. Patanè, *Real-Time Processing of Ultrasound Images*, Synthesis Lectures on Visual Computing: Computer Graphics, Animation, Computational Photography and Imaging, https://doi.org/10.1007/978-3-032-05805-8_4

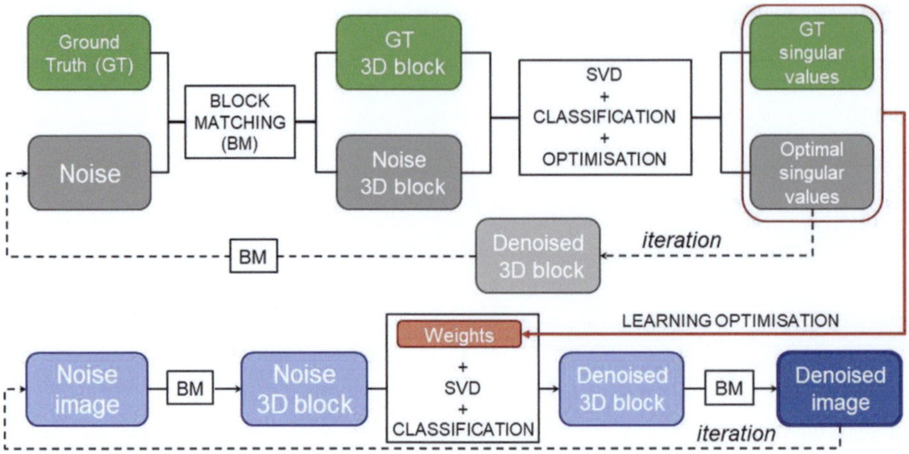

Fig. 4.1 Learning-based for denoising: pipeline. Image courtesy of [2]

We present a background on the low-rank methods and SVD (Sect. 4.1) and a description of the learning-based low-rank method (Sect. 4.2).

4.1 Background on Low-Rank Methods and SVD

Given a noisy signal $\mathbf{Y}$, denoising methods want to estimate the ground truth signal $\mathbf{X}$ with respect to the noisy component $\mathbf{N}$, as $\mathbf{Y} = f(\mathbf{X}, \mathbf{N})$ where f is a generic operator (e.g., additive, multiplicative). Both the signals and the noise can be defined as 2D matrices $m \times n$ or as continuous functions $\mathbf{X}, \mathbf{Y}, \mathbf{N} : \Omega \subset \mathbb{R}^2 \rightarrow \mathbb{R}$. In this context, a denoising method aims to find an approximation of the noisy signal that penalises undesirable solutions through regularisation.

$$\hat{\mathbf{X}} = \arg\min_{\mathbf{X}} \left\{ \|\mathbf{Y} - \mathbf{X}\|^2 + \lambda P(\mathbf{X}) \right\}, \tag{4.1}$$

where $\|\mathbf{Y} - \mathbf{X}\|^2$ is the discrete sum/integral over the domain of the squared difference of the noisy and ground truth signal in the discrete and continuous context, respectively. The $P(\mathbf{X})$ denotes a penalty function which is λ-weighted and is higher as the solution recedes from the target one, according to the desired properties. In this context, several formulations describe different classes of representation and regularisation.

In sparse-representation models (e.g., Fourier transform, Wavelets), the predicted signal is represented as a sparse representation $\mathbf{\Theta}$ of a dictionary of basis functions $\mathbf{W}$. The minimisation problem in Eq. (4.1) finds the minimum number of bases to represent the noisy signal, and the penalisation expresses certain properties of the dictionary-based representation, e.g., the reduction of high-frequency components of the Fourier transformation, as

$$\hat{\mathbf{\Theta}} = \arg\min_{\mathbf{\Theta}} \left\{ \|\mathbf{Y} - \mathbf{W\Theta}\|^2 + \lambda P(\mathbf{\Theta}) \right\}.$$

The total variation method is a variational model

$$\hat{\mathbf{X}} = \arg\max_{\mathbf{X}} \left\{ \int_{\Omega} (\mathbf{Y(t)} - \mathbf{X(t)})^2 + \lambda P(\mathbf{X}) \right\},$$

where the penalisation is the integral of the norm of the gradient of the solution $P(\mathbf{X}) = \int_{\Omega} |\nabla \mathbf{X(t)}| d\mathbf{t}$. The *nuclear norm minimisation* (NNM) model penalises the nuclear norm, defined as the sum of the singular values of a matrix. $\|\mathbf{X}\|_* = \sum_i |s_i(\mathbf{X})|$ as

$$\hat{\mathbf{X}} = \arg\min_{\mathbf{X}} \left\{ \|\mathbf{Y} - \mathbf{X}\|^2 + \lambda \, \|\mathbf{X}\|_* \right\}.$$

NNM can be equivalently solved as a low-rank problem by soft-thresholding the variable matrix's singular values through a fixed quantity λ, as shown in [1]. The NNM is further generalised to the Weighted Nuclear Norm Minimisation (WNNM, [10]), where different threshold values are applied to each singular value as

$$\hat{\mathbf{X}} = \arg\min_{\mathbf{X}} \left\{ \|\mathbf{Y} - \mathbf{X}\|^2 + \|\mathbf{X}\|_{w,*} \right\}, \tag{4.2}$$

where $\|\mathbf{X}\|_{w,*} = \sum_i |w_i s_i(\mathbf{X})|$, with $w_i \geq 0$ is a non-negative weight assigned to $s_i(\mathbf{X})$. In this context, we discuss some numerical and computational properties of a matrix's singular value decomposition (SVD).

SVD: Numerical Properties

Every matrix $\mathbf{A} \in C^{m \times n}$ has a singular values decomposition, which is the factorisation of $\mathbf{A}$ into the product of three matrices $\mathbf{A} = \mathbf{USV}^T$. $\mathbf{U}$ is $m \times m$ orthogonal matrix, whose columns $u_1, \ldots, u_m$ are m-dimensional orthonormal left singular vectors of $\mathbf{A}$; $\mathbf{V}$ is $n \times n$ orthogonal matrix, whose columns $v_1, \ldots, v_n$ are n-dimensional orthonormal right singular vectors of $\mathbf{A}$; $\mathbf{S}$ is an $m \times n$ diagonal matrix with non-negative entries sorted in decrescent order. The singular values are uniquely defined; if the singular values are all distinct, then the right and left singular vectors are unique. On the contrary, when some sets of singular values are equal, the corresponding singular vectors span some subspace, and any set of orthonormal vectors spanning this subspace can be used as the singular vectors.

The columns and the rows of $\mathbf{U}$ and $\mathbf{V}$ are computed as the eigenvectors of $\mathbf{AA}^T$ and $\mathbf{A}^T\mathbf{A}$ respectively, while the diagonal elements s_i of $\mathbf{S}$ are the square roots of the n eigenvalues of $\mathbf{A}^T\mathbf{A}$, with $s_1 \geq s_2 \geq \cdots \geq s_n$. Moreover, the 2-norm and the Frobenius norm of $\mathbf{A}$ are $\|\mathbf{A}\|_2 = s_1$ and $\|\mathbf{A}\|_F = \sqrt{s_1^2 + \cdots + s_r^2}$ respectively, where r is the rank of $\mathbf{A}$. The image space and the null space of $\mathbf{A}$ can be computed as $range(\mathbf{A}) =< \mathbf{u}_1, \ldots, \mathbf{u}_r >$ and $null(\mathbf{A}) =< \mathbf{v}_{r+1}, \ldots, \mathbf{v}_n >$. Geometrically, a $m \times n$ matrix is an application from $\mathbb{R}^n$

to $\mathbb{R}^m$ that performs a rotation in the domain (multiplication by $\mathbf{V}^T$), followed by scaling plus adding or deleting dimensions (multiplication by $\mathbf{S}$), followed by a rotation in the range (multiplication by $\mathbf{U}$).

The matrix $\mathbf{A}$ of rank r can be approximated through a low-rank matrix of rank k, as $\mathbf{A}_k = \sum_{i=1}^{k} s_i \cdot \mathbf{u}_i \mathbf{v}_i^T$. This is equivalent to $\mathbf{A}_k = \mathbf{U}_k \mathbf{S}_k \mathbf{V}_k^T$, where $\mathbf{U}_k$ and $\mathbf{V}_k$ are the first k columns of $\mathbf{U}$ and $\mathbf{V}$ respectively, and $\mathbf{S}_k$ are the first k singular values, i.e., the first k elements of the diagonal of $\mathbf{S}$. $\mathbf{A}_k$ is the best approximation of rank k of the matrix $\mathbf{A}$. In fact, for every $m \times n$ matrix $\mathbf{A}$, rank target $k \geq 1$, and $m \times n$ matrix $\mathbf{B}$ of rank k, $\|\mathbf{A} - \mathbf{A}_k\|_F \leq \|\mathbf{A} - \mathbf{B}\|_F$, where $\mathbf{A}_k$ is the rank-k approximation derived from the SVD of $\mathbf{A}$.

Furthermore, we estimate the approximation error of $\mathbf{A}_k$ with respect to $\mathbf{A}$ as $\|\mathbf{A} - \mathbf{A}_k\|_2 = s_{k+1}$ and $\|\mathbf{A} - \mathbf{A}_k\|_F = \sqrt{s_{k+1}^2 + \cdots + s_r^2}$. The low-rank approximation $\mathbf{A}_k$ requires $O(k(m+n))$ instead of $O(m \cdot n)$ memory. Finally, the solution of Eq. (4.2) can be computed through the weighted threshold of the SVD decomposition as $\hat{\mathbf{X}} = \mathbf{U}\hat{\mathbf{S}}\mathbf{V}$, where

$$\hat{\mathbf{S}} = \arg\min_{\mathbf{S}^*} \left\{ \|\mathbf{S} - \mathbf{S}^*\|_F^2 + \|\mathbf{S}^*\|_{w,*} \right\}.$$

SVD: Computational Properties

Most of the algorithms developed for the computation of the SVD find square roots of eigenvalues of $\mathbf{A}^T\mathbf{A}$ without actually computing the matrix-matrix multiplication, which is a computationally expensive operation. The reduction of the input matrix to a bidiagonal form through the Householder transformation [12] allows several algorithms to compute the SVD through bisection [6], divide and conquer [9], and QR factorisation [7] methods. Other algorithms directly apply to the input matrix through diagonalising rotations, a typology of orthogonal transformations that keeps track of a matrix's eigenvalues, eigenvectors, singular values, and singular vectors, as well as angles and lengths. The application of a sequence of rotations $\mathbf{A}^k$ converges to the diagonal matrix of the singular values, i.e., $\lim_{k\to\infty} \mathbf{A}^k = \mathbf{S}$; in particular, the Jacobi rotation annihilates a symmetric pair of off-diagonal entries. In [15], the analogous of the Jacobi method for symmetric matrices is applied. The application of the cyclic Jacobi method [8] consists of an iteration of sweeps which does not require the determination of the largest off-diagonal element of $\mathbf{A}^k$ and an orthogonal reduction of the $m \times n$ input matrix generates an output matrix where all the non-zeros are in the upper $n \times n$ portion. The *Hestens-Jacobi* method [11] accounts for the equivalence between the orthogonalisation of two vectors and the annihilation of a matrix element using orthogonal plane rotations.

Recently, the development of methods for the SVD accounts for parallel computation, favouring those algorithms that allow a distribution of the operations among different processes. In [19], a slight increase in the computational cost of the method by sacrificing the convergence rate allows the algorithm to distribute the calculation over several processes with benefits in terms of computational cost. For further details on the SVD decomposition and its application for low-rank problems, we refer the readers to [3, 14, 17].

4.2 Learning-Based Low-Rank Denoising Method

We discuss the optimisation model (Sect. 4.2.1), the learning model (Sect. 4.2.2), and the experimental results (Sect. 4.2.3).

4.2.1 Optimisation Model for the Computation of the Optimal Thresholds

An input image $\mathbf{Y} = f(\mathbf{X}, \mathbf{N})$ is composed of the ground truth $\mathbf{X}$ and the noisy component $\mathbf{N}$, where f defines the combination of one or more types of noise (e.g., additive, impulsive, multiplicative). Low-rank approximations recover the underlying low-rank matrix from its degraded observation. In this class, the *nuclear norm minimisation* recovers an estimation $\hat{\mathbf{X}}$ of the ground truth signal through the minimisation of the energy functional $\|\mathbf{Y} - \mathbf{X}\|_F^2 + \lambda\|\mathbf{X}\|_*$, where the *nuclear norm* $\|\mathbf{X}\|_*$ is the sum of the singular values of $\mathbf{X}$ and measures the compliance of the approximated image to the high-frequency components of the input image. This problem is equivalent to recover the approximated image $\hat{\mathbf{X}} = \mathbf{U}(\tau(\mathbf{S}))\mathbf{V}^\top$, where $\mathbf{Y} = \mathbf{U}\mathbf{S}\mathbf{V}^\top$, $\mathbf{S} := \text{diag}(\mathbf{s})$, is the SVD of the noisy signal, and $\tau(\mathbf{S}) = \text{diag}(\mathbf{s} - \boldsymbol{\lambda})$ with $\boldsymbol{\lambda}$ constant vector. The thresholding $\tau(\mathbf{S})$ improves the image reconstruction by removing the noise component while preserving image features. Weighted thresholds imply a different shrinkage to each singular value, as $\mathbf{S}_{hh} - \boldsymbol{\lambda}_h$, where $\mathbf{S}_{hh}$ is the (h, h) entry of $\mathbf{S}$. The learning-based method trains a network to predict the optimal thresholds $\boldsymbol{\lambda}_h$.

Given a $n \times n$ ground truth image $\mathbf{X}$ and a noisy instance $\mathbf{Y} = f(\mathbf{X}, \mathbf{N})$, the optimal thresholds $\boldsymbol{\lambda}_h$ are computed through the minimisation of the distance between the ground truth and the reconstructed signal $\hat{\mathbf{X}}$, as

$$E(\boldsymbol{\lambda}) = \sum_{i=1}^{n}\sum_{j=1}^{n} |\mathbf{X}_{ij} - \hat{\mathbf{X}}_{ij}|^2, \tag{4.3}$$

where the reconstructed signal is computed through the SVD of the $\mathbf{Y}$ signal with weighted thresholds, i.e., $\hat{\mathbf{X}}_{ij} = \sum_{h=1}^{n} \mathbf{U}_{ih}(\mathbf{S}_{hh} - \boldsymbol{\lambda}_h)\mathbf{V}_{hj}$. The first-order derivatives of the energy functional $E(\boldsymbol{\lambda})$ concerning the variables of the optimisation problem $\boldsymbol{\lambda}$ are

$$\frac{\partial E(\boldsymbol{\lambda})}{\partial \boldsymbol{\lambda}_k} = 2\sum_{i,j=1}^{n} \left[\mathbf{X}_{ij} - \sum_{h=1}^{n} \mathbf{U}_{ih}(\mathbf{S}_{hh} - \boldsymbol{\lambda}_h)\mathbf{V}_{hj}\right] \mathbf{U}_{ik}\mathbf{V}_{kj}.$$

The minimum of Eq. (4.3) is computed through the iterative optimisation method L-BFGS (*Limited-memory Broyden, Fletcher, Goldfarb, Shanno*) [22], which finds the roots of the derivative of the energy functional. L-BFGS is an optimisation algorithm in the family of quasi-Newton methods that approximates the Broyden-Fletcher-Goldfarb-Shanno algorithm (BFGS) using limited computer memory. Analogously to BFGS, the L-BFGS solver estimates the inverse Hessian matrix for the minimum search in the variable space; however, the L-BFGS method represents the approximation implicitly through a few vectors, thus involving a limited memory requirement. At each iteration, a small history of the past updates of the variables ($\boldsymbol{\lambda}$) and the gradient of the energy functional $E(\cdot)$ in Eq. (4.3) is used to identify the direction of steepest descent and to implicitly perform operations requiring vector products with the inverse Hessian matrix. For the L-BFGS method, the memory storage is $O(u^2)$ and the computational cost is $O(uv)$ at each iteration, where u is the total number of variables, and v is the number of steps stored in memory.

Application of Optimal Thresholds for Denoising

The direct application of the optimal thresholds to the noisy input image has reduced accuracy due to the complexity of the geometries of the image and the difficulty in distinguishing the edges and structures from the noise components. Therefore, the introduction of a *block-matching* algorithm allows the user to compute stacks of similar patches with redundancy properties and apply the SVD and the optimal thresholds to these stacks, where the LLR method can identify and then remove the noisy component, maintaining the geometric and texture structures. Furthermore, applying the optimal thresholds through an *iterative approach* allows for progressively improving the accuracy of the solution.

Block-matching Given a pixel y, the *patch* P_y is the set of pixels in the neighbourhood of y. The *search window* is the set of selected patches used to search for the closest ones to a reference patch, according to certain metrics. The *stack*, or *3D block*, is the set of patches similar to a reference patch. The block-matching algorithm stacks similar patches into a 3D structure and uses the redundancy of the stack to remove the noise. Then, the patches are aggregated by a collaborative filter to compose the denoised image.

Iterative approach The method is applied for 20 iterations: at each iteration, the denoised image of the previous iteration is used as a noisy image, while the ground truth is fixed. The 3D blocks are computed with the block-matching algorithm, and their ground truth counterparts are extracted from the ground truth image. Then, the optimal thresholds $\boldsymbol{\lambda}_h$ for the SVD are computed through the minimisation of Eq. (4.3) for each 3D block, and the denoised image is reconstructed.

4.2.2 Learning Model: Setting and Optimisation

The data set and the learning model of LLR for the computation of the optimal thresholds are described.

Training and Clustering Data Set

Applying an arbitrary noise to a set of input images, a data set of ground truth and noisy images is built to train a learning-based method to predict the optimal thresholds. In particular, the input of the network is the vector of the singular values of the noisy block, and the target is the vector of the optimal thresholds (Eq. (4.3)). Instead of defining a single data set for all the images, the images are organised according to four parameters: (i) *type of noise*, (ii) *iteration*, (iii) *noise intensity*, and (iv) *singular values clustering*. This classification improves the network's specificity and the prediction accuracy of the thresholds. The four parameters are nested: for each image noise (i) and iteration step, (ii) the 3D block's noise intensity, and (iii) is computed with the *nonlinear noise estimator* (NOLSE) [16] metric and c intervals are defined so that the NOLSE values are uniformly distributed into c classes. The singular values (iv) are clustered into k groups with the *k-means* algorithm; then, a Support Vector Machine (SVM) [4] is trained to classify the vectors of singular values according to the *k-means* labels. For each combination of the four parameters, the matrix $\mathbf{P} = \{\mathbf{P}_k\}_{k=1}^{N}$ is defined with $\mathbf{P}_k = [p_1, \ldots, p_n]^\top$, where N is the number of 3D blocks and n is the lower dimension of the 3D block; the p elements are the singular values computed through the SVD. These values are associated with the corresponding optimal thresholds $\mathbf{Q} = \{\mathbf{Q}_k\}_{k=1}^{N}$, computed as the minimisation of Eq. (4.3). The trained network is applied to predict the optimal denoising for US images (Sect. 4.2.3); in this context, learning models trained on images with speckle noise at different intensities are used; then, five iterations are applied with the respective trained network, computing the noise intensity and k-means prediction for each block.

Optimisation Model

A dedicated learning-based network is computed for each combination of the four parameters (Sect. 4.2.2). Given the singular values of a 3D block, a network predicts the thresholds that reconstruct the best approximation of the ground truth block. The network weights are defined through the matrix $\mathbf{W}$ of dimension $n \times n$ and are computed by minimising the distance between the target and the predicted thresholds as

$$F(\mathbf{W}) = \sum_{k=1}^{N} \sum_{j=1}^{n} \left| \mathbf{Q}_{jk} - \hat{\mathbf{Q}}_{jk} \right|^2, \tag{4.4}$$

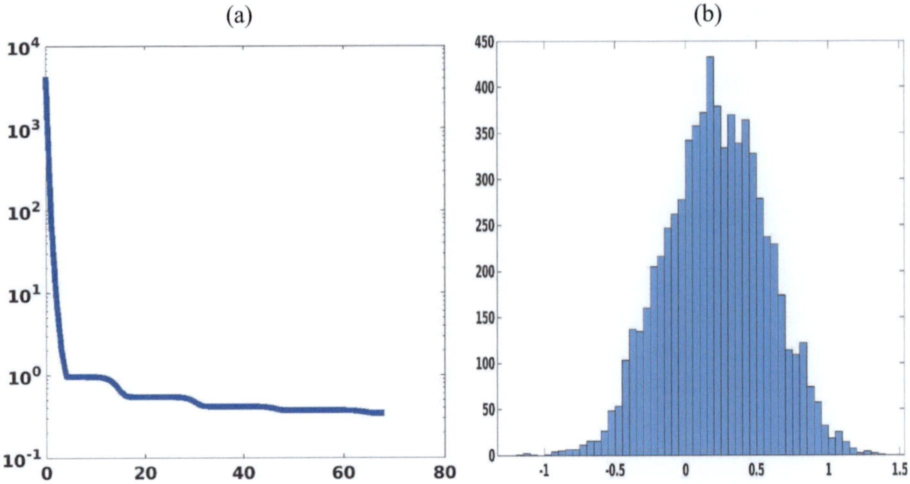

Fig. 4.2 **a** Minimisation of the functional (Eq. 4.4) value (y-axis) normalised on the number of vectors $N = 200$, with respect to the number of iterations (x-axis); **b** frequency (y-axis) of the difference between the optimal and the predicted thresholds (x-axis) of the first singular value of the SVD, for all the blocks of the cameraman image. Image courtesy of [2]

where $\hat{\mathbf{Q}}_{jk} = \sum_{i=1}^{n} \mathbf{W}_{ij}\mathbf{P}_{ik}$ are the predicted thresholds. The first-order derivatives of Eq. (4.4) are

$$\frac{\partial F(\mathbf{W})}{\partial \mathbf{W}_{lm}} = -2\sum_{k=1}^{N}\left(\mathbf{Q}_{mk} - \sum_{i=1}^{n}\mathbf{P}_{ik}\mathbf{W}_{im}\right)\mathbf{P}_{lk}, \tag{4.5}$$

and the second-order derivatives are

$$\frac{\partial^2 F(\mathbf{W})}{\partial^2 \mathbf{W}_{lm}} = 2\sum_{k=1}^{N}|\mathbf{P}_{lk}|^2.$$

Since the functional in Eq. (4.4) is quadratic and its second-order derivatives are always strictly positive, the functional $F(\cdot)$ is strictly convex, and it has a unique minimum.

Figure 4.2a shows the convergence of the minimisation to the optimal solution computed with the L-BFGS algorithm (Sect. 4.2.1). The solver takes 67 iterations to converge to the minimum of the functional. Figure 4.2b shows the frequency of the error between the optimal and the predicted thresholds for the first singular value $\#_i\{x \in (\boldsymbol{Q}_{1k} - \hat{\boldsymbol{Q}}_{1k}) : h_i \leq x \leq h_{i+1}\}$, with h bins of the histogram. The average error is 0.3, and the upper and lower bounds of the error are ± 1.5.

The minimisation of the functional in Eq. (4.4) is equivalent to the solution of the linear system $\nabla F(\mathbf{W}) = (\mathbf{P}\mathbf{P}^\top)\mathbf{W} - \mathbf{P}\mathbf{Q}^\top = \mathbf{0}$ (c.f., Eq. (4.5)), with $n \times n$ coefficient matrix $\mathbf{P}\mathbf{P}^\top$ and $rank(\mathbf{P}\mathbf{P}^\top) = n$ when N is sufficiently greater than n. Increasing the number N of vectors of singular values (i.e., the columns of $\mathbf{P}$ and $\mathbf{Q}$) from 250 to 2500 (Fig. 4.3a), the

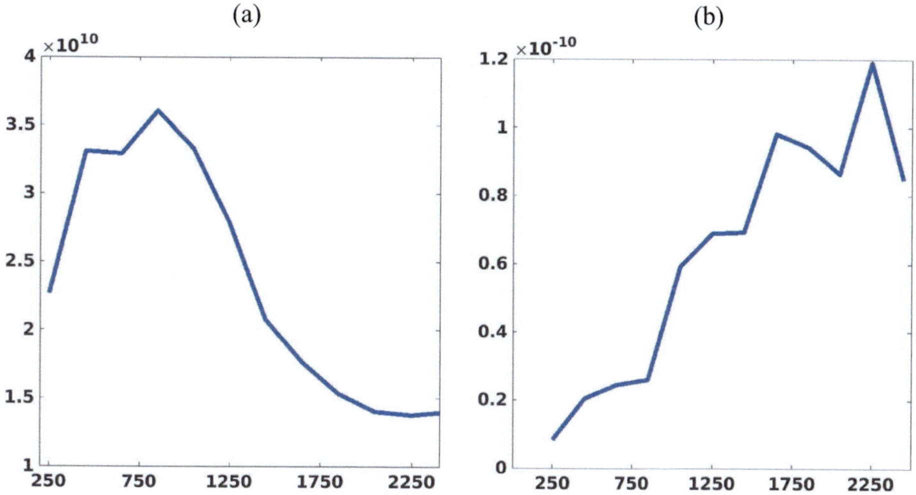

Fig. 4.3 **a** Conditioning of the coefficient matrix (y-axis) with respect to N (x-axis). **b** Approximation error of the linear system (y-axis) with respect to N (x-axis). Image courtesy of [2]

condition number of the coefficient matrix passes from 2.3×10^{10} to 1.5×10^{10}; given the solution $\widetilde{\mathbf{W}}$, the approximation error of the linear system $\|\nabla F(\widetilde{\mathbf{W}})\|_F/N$ passes from 8.3×10^{-12} to 8.4×10^{-11} (Fig. 4.3b).

4.2.3 Experimental Results

Training Results

LLR selects a data set of 39 images [21] of different resolutions and subjects. LLR applies an artificial noise to the images (e.g., Gaussian noise with mean 0 and variance 0.01) and then computes the 3D blocks with block-matching. LLR computes the input singular values and the target thresholds. Images are classified according to the noise type (e.g., speckle, Gaussian) and intensity (e.g., $\sigma = 0.05$ in speckle noise) of the input image. Then, LLR splits the data into $c = 5$ 3D block noise intensity classes and $k = 3$ k-means labels for a total of 15 instances for each iteration. The final data set is composed of a set of 15 matrices $n \times N$, where N is the total number of 3D blocks, and n is the number of singular values (i.e., 49 in our tests, corresponding to a 7×7 patch). This approach is repeated at each iteration: for the computation of the data set at iteration t, the block matching is applied for $(t-1)$ iterations. At each iteration, the proper networks previously computed by the minimisation model are used to predict the thresholds and denoise the related 3D block. At iteration t, the optimal thresholds are computed through an optimisation step (Eq. (4.3)). In the experimental tests,

5 iterations of the training are applied for a total number of 75 networks (i.e., $5 \times 3 \times 5$) for each type of noise and noise intensity.

Computation Time The computation of the input/optimal thresholds (Eq. (4.3)) and of the network weights is performed offline. For the computation of the training data set, LLR uses 39 images for a total number of 257,240 pairs of input/target vectors for each noise type and iteration. The execution time for the generation of the data set of each noise type/intensity is approximately 120 minutes for each iteration. For the computation of the weights, the variables of each network are 49×49 matrices; LLR considers 3 types of noise (i.e., salt and pepper, speckle, Gaussian), 5 iterations for each noise, 5 3D blocks' noise intensity classes, and three $k - means$ labels, for a total number of 225 networks. The execution time of the training of each network is about 15 minutes. Tests have been executed with MATLAB R2020a on a workstation with 2 Intel i9-9900KF CPUs (3.60GHz) and 32GB RAM. Additional tests with increased noise intensity (i.e., Gaussian and speckle) and different types of noise (i.e., Poisson and exponential) are discussed in the additional material.

Denoising Results

We discuss the generality of the LLR method from the type of noise applied. Applying a speckle noise with $\sigma = 0.05$, the LLR method improves previous work and shows better preservation of the main features and details and good noise removal. Furthermore, the LLR method does not introduce blurred patterns. In comparison, WNNM removes a larger amount of noise while losing several details and increasing the blurring effect [2]. Table 4.1 summarises the quantitative metrics of the LLR method and previous work. With Gaussian noise, the LLR method has comparable results with state-of-the-art methods, which are specific for this type of noise. In (c), the PSNR value is 26.01 for the LLR method, 26.13 for WNNM (best result), and lower than 26 for BM3D and NCSR. Instead, the LLR method

Table 4.1 We report the PSNR metric of LLR denoising, compared with previous work: (a) exponential, $\mu = 0.02$, (b) speckle, $\sigma = 0.10$, (c) Poisson, and (d) Gaussian noise, $\mu = 0, \sigma^2 = 0.02$.

PSNR	(a)	(b)	(c)	(d)
DUBD	31.79	16.01	28.11	17.35
SHEARLAB	23.40	27.89	26.85	25.82
ACVA	32.95	25.25	34.11	28.12
BM3D	32.18	25.56	33.98	27.98
WNNM	32.08	28.29	33.87	**28.45**
NCSR	32.20	28.37	33.30	27.15
LLR	**38.82**	**29.18**	**34.43**	28.18

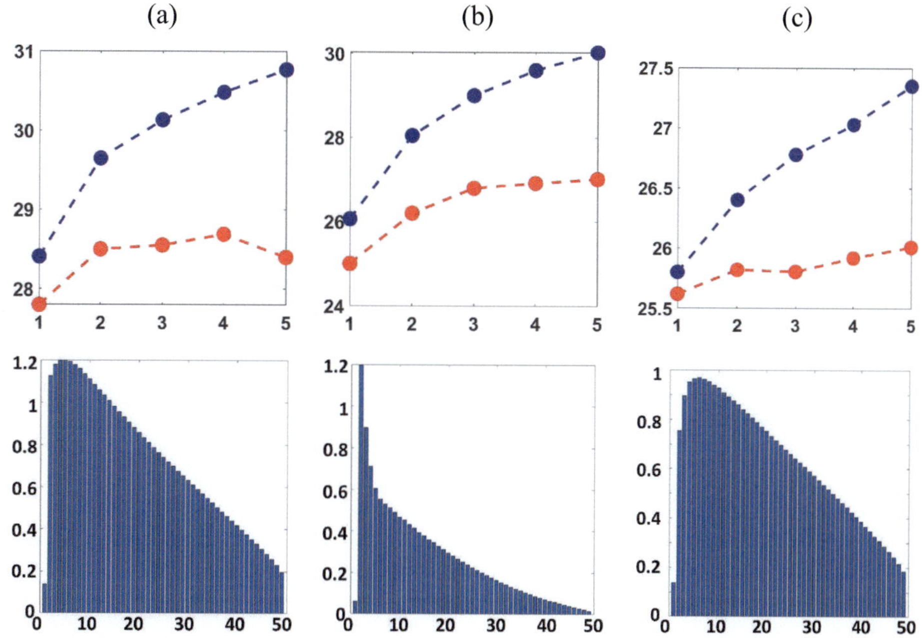

Fig. 4.4 1st Row: PSNR (y-axis) comparison of the number of iterations (x-axis) between optimal values (blue), and LLR method's values (red). 2nd Row: average optimal thresholds (y-axis) of each singular value (x-axis). **a** Cameraman (256×256), speckle noise ($\sigma = 0.05$); **b** boat (256×256), salt and pepper noise (density 0.05); **c** mandrill, Gaussian noise ($\mu = 0, \sigma = 0.01$). Image courtesy of [2]

has better results when other types of noise (e.g., salt and pepper, speckle) are applied. In Fig. 4.4, we compare the PSNR of the LLR method with the optimal PSNR results; in (a), the LLR method has a PSNR value of 28.69 after 4 iterations, while the optimal PSNR value is 30.53. The error of the network prediction of the thresholds, even if small, leads the LLR method to diverge from the optimal reconstruction. After five/six iterations, the LLR method is not able to further improve the image approximation. Considering a large data set [18] of images of different resolutions and subjects and applying different types of artificial noise, the LLR method has slightly better results than previous work in the case of Gaussian noise while it widely improves the PSNR values in the case of speckle and salt and pepper noise (Fig. 4.5).

Computation Time The execution time for denoising each 3D block is 0.015 s; this time is composed of the noise intensity computation through the NOLSE metric (66%), the SVD (5%), the SVM prediction of the label of the singular values (14%), and the network prediction of the optimal thresholds (15%). Due to the operations performed for each 3D block, the LLR method (330 s) increases the execution time for the denoising of a 256×256 image, with respect to previous work: BM3D (1 sec.), WNNM (100 s), and NCSR (180 s).

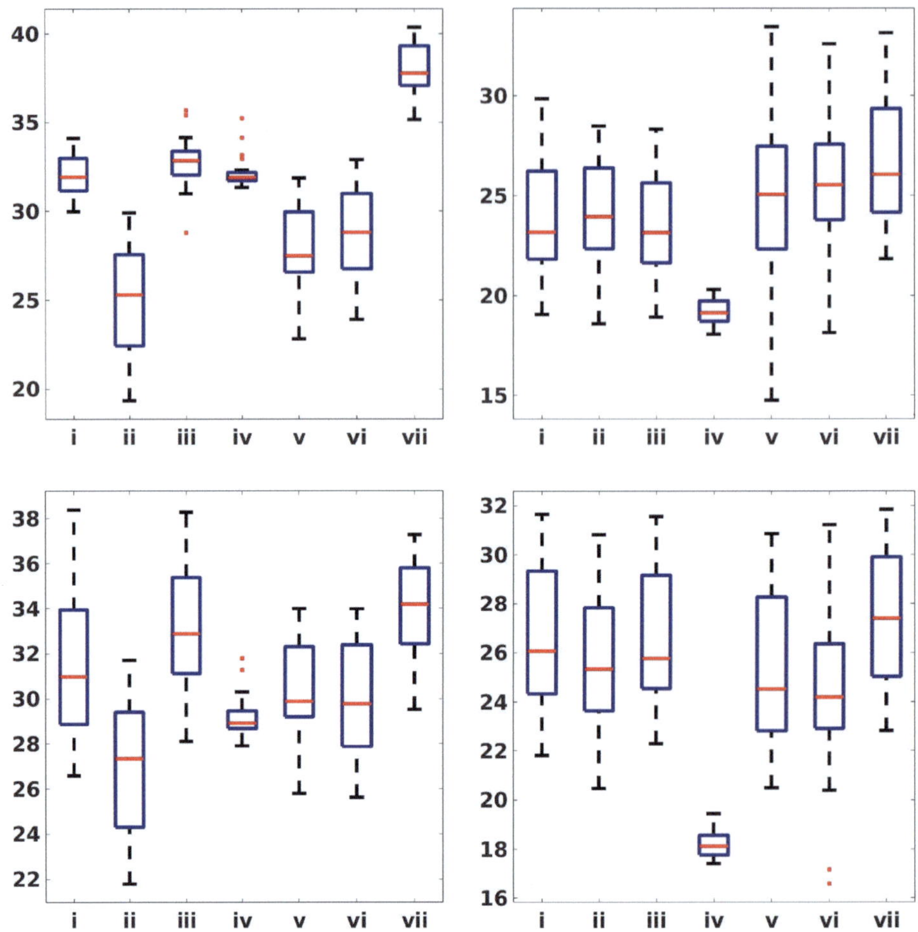

Fig. 4.5 PSNR (y-axis) average values of LLR and state-of-the-art methods (x-axis) with exponential $\mu = 0.02$ (top-left), speckle $\sigma = 0.10$ (top-right), Poisson (bottom-left), Gaussian $\mu = 0$, $\sigma^2 = 0.02$ (bottom-right) noise. (i) BM3D, (ii) SHEARLAB, (iii) ACVA, (iv) DUBD, (v) WNNM, (vi) NCSR, (vii) LLR. Image courtesy of [2]

Selected Thresholds

The training data set of the Gaussian noise with $\mu = 0$ and $\sigma^2 = 0.01$ is composed of more than 257K elements, each one composed of a couple of vectors (i.e., input and optimal target) in $\mathbb{R}^{49}$. Figure 4.6a shows the partition of these elements, according to the 3D block noise intensity and the k-means classification. Since each 3D block noise intensity class contains the same number of elements, we have a regular partition of the training dataset. In contrast, the k-means classification does not provide a uniform partition of the elements; indeed, some networks are trained with more data than others. However, each network has

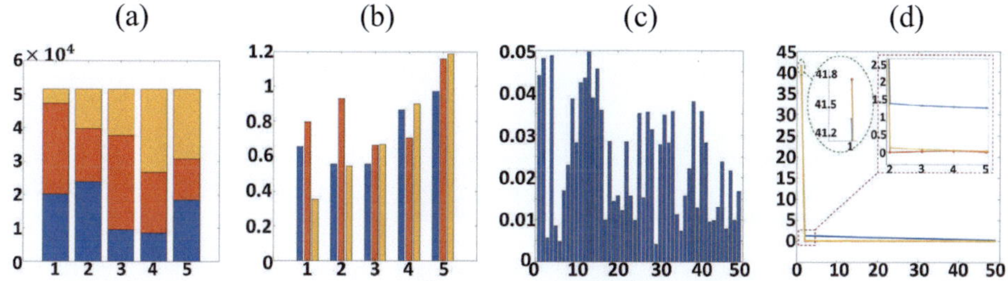

Fig. 4.6 **a** Partition of the elements (i.e., input and target vectors) for the training. Each noise intensity class (x-axis) is divided into 3 k-means classes (blue, orange, and yellow); y-axis shows the cumulated number of elements of each class. **b** Input (blue), optimal (orange), and predicted (yellow) singular values (y-axis) of the cameraman image with speckle noise for the 49 (x-axis) singular values. **c** Prediction error (y-axis) of the 49 singular values (x-axis) of a 3D block of the cameraman image with speckle noise. **d** Partition of the prediction error (y-axis), when varying the noise intensity classes (x-axis): each class is composed of three SVM-predicted classes (i.e., blue, orange, and yellow). Image courtesy of [2]

a sufficient number of elements due to the huge data set; in fact, the minimum number of elements is $N = 4204$, related to 3D block noise class $c = 1$ and k-means label $k = 3$.

Figure 4.6b shows a comparison among the input, optimal, and predicted singular values of a specific 3D block. The optimal singular values reflect the thresholds in Fig. 4.4 (2nd col.); for example, the shrinkage of the second singular value is larger than the first one. The predicted singular values are very close to the optimal singular values; in fact, the error for the prediction of a 49-dimensional vector is 0.6, which means an average error of 0.01 for each element of the vector. Analysing the error between the prediction and the optimal singular values $(\mathbf{Q}_{jk} - \hat{\mathbf{Q}}_{jk})_{j=1}^{49}$, the first values show a higher prediction error than the last ones (Fig. 4.6c), due to the different magnitude and variability of these values. However, there is no clear correlation between these factors. The correlation between the prediction error of the thresholds and the data clustering shows that the error increases with the 3D block noise intensity (Fig. 4.6d), as the noise intensity implies a more irregular distribution of the singular values, which is harder to train and predict. Instead, the SVM-predicted classes and the prediction error of the network do not show any correlation.

Experimental Results on US Images

The LLR denoising method is compared with previous work on ultrasound images from the different anatomical districts (Fig. 4.7). In particular, LLR models are trained on images with speckle noise at different intensities: $\sigma = 0.05$ (Fig. 4.7, LLR (i)), and $\sigma = 0.10$ (Fig. 4.7, LLR (ii)). The learning model trained with images with higher noise applies larger thresholds to the singular values of the SVD, resulting in more intense smoothing. The LLR method correctly preserves the edges of the obstetric district (Fig. 4.7a) and the mass of the

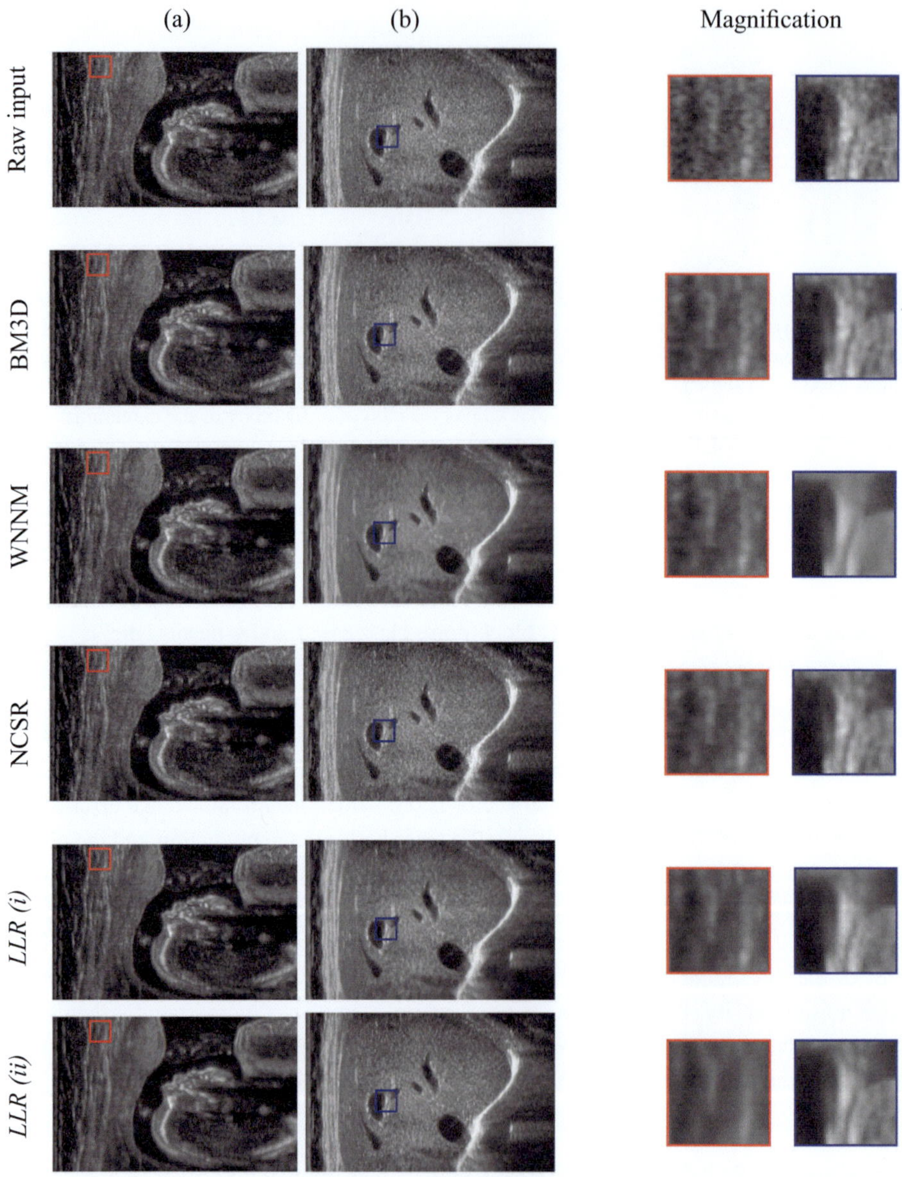

Fig. 4.7 Comparison between previous work and LLR denoising with low (i) and medium (ii) smoothing intensity. Image courtesy of [2]

abdominal districts (Fig. 4.7b) while removing the noise component; this result is perceptible in the magnification, comparing the LLR method with previous work.

The execution time for denoising each 3D block is 0.015 s; this time is composed of the noise intensity computation through the NOLSE metric (66%), the SVD (5%), the SVM prediction of the label of the singular values (14%), and the network prediction of the optimal thresholds (15%). Due to the operations performed for each 3D block, the LLR method (330 s) increases the execution time for the denoising of a 256×256 image, compared to previous work: BM3D (1 s), WNNM (100 s), and NCSR (180 s). This execution time does not satisfy the real-time constraint for the Ultrasound application.

References

1. J.-F. Cai, E.J. Candès, Z. Shen, A singular value thresholding algorithm for matrix completion. SIAM J. Optim. **20**(4), 1956–1982 (2010)
2. S. Cammarasana, G. Patane, Learning-based low-rank denoising, in *Signal, Image and Video Processing* (2022), pp. 1–7
3. A.K. Cline, I.S. Dhillon, *Computation of the Singular Value Decomposition* (CRC Press, 2006)
4. N. Cristianini, J. Shawe-Taylor, *An Introduction to Support Vector Machines and Other Kernel-Based Learning Methods* (Cambridge University Press, 2000)
5. K. Dabov et al., Image denoising with block-matching and 3D filtering, in *Image Processing: Algorithms and Systems, Neural Networks, and Machine Learning*, vol. 6064 (SPIE, 2006), p. 606414
6. J. Demmel, W. Kahan, Accurate singular values of bidiagonal matrices. SIAM J. Sci. Stat. Comput. **11**(5), 873–912 (1990)
7. J.W. Demmel, *Applied Numerical Linear Algebra* (SIAM, 1997)
8. G.E. Forsythe, P. Henrici, The cyclic Jacobi method for computing the principal values of a complex matrix. Trans. Amer. Math. Soc. **94**(1), 1–23 (1960)
9. M. Gu, S.C. Eisenstat, A divide-and-conquer algorithm for the bidiagonal SVD. SIAM J. Matrix Anal. Appl. **16**(1), 79–92 (1995)
10. S. Gu et al., Weighted nuclear norm minimization with application to image denoising, in *Proceedings of the IEEE Conference on Computer Vision and Pattern Recognition* (2014), pp. 2862–2869
11. M.R. Hestenes, Inversion of matrices by biorthogonalization and related results. J. Soc. Ind. Appl. Math. **6**(1), 51–90 (1958)
12. A.S. Householder, Unitary triangularization of a nonsymmetric matrix. J. ACM **5**(4), 339–342 (1958)
13. Z. Hu et al., A comprehensive survey for low rank regularization (2018). arXiv: 1808.04521
14. V. Klema, A. Laub, The singular value decomposition: its computation and some applications. IEEE Trans. Autom. Control **25**(2), 164–176 (1980)
15. E.G. Kogbetliantz, Solution of linear equations by diagonalization of coefficients matrix. Q. Appl. Math. **13**(2), 123–132 (1955)
16. O. Laligant, F. Truchetet, E. Fauvet, Noise estimation from digital step-model signal. IEEE Trans. Image Proc. **22**(12), 5158–5167 (2013)
17. T. Roughgarden, G. Valiant, CS168: The Modern Algorithmic Toolbox Lecture# 9: the Singular Value Decomposition (SVD) and Low-Rank Matrix Approximations (2015). http://theory.stanford.edu/tim/s15/l/l9.pdf. Accessed 18 Sep 2022
18. P. Roy et al., Effects of degradations on deep neural network architectures (2018). arXiv:1807.10108

19. V. Strumpen, H. Hoffmann, A. Agarwal, A stream algorithm for the SVD, in *Technical Memo*, MIT-LCS-TM-641 (2003), pp. 1–29
20. C. Tian et al., Deep learning on image denoising: an overview. Neural Netw. **131**, 251–275 (2020)
21. A.G. Weber, The USC-SIPI image database version 5. USC-SIPI Rep. **315**(1) (1997)
22. C. Zhu et al., Algorithm 778: L-BFGS-B: Fortran subroutines for largescale bound-constrained optimization. ACM Trans. Math. Softw. **23**(4), 550–560 (1997)

Real-Time Denoising of 2D US Images 5

Two strategies achieve real-time computation and maintain the good results of the best denoising methods in noise removal and edge preservation. After the identification of the denoising method that best fits the requirements for ultrasound images: (i) the development of a computationally optimised version of a selected denoising algorithm, exploiting HPC tools, CPUs and GPUs, and low-level programming languages; (ii) the design and implementation of a deep learning framework that learns to replicate the output of the selected denoising algorithm.

Ultrasound images can be acquired at different planes through the manual/automatic movement of the probe; furthermore, the physician can acquire both static and dynamic images. These acquisitions are associated with different input signals: 2D images, 2D videos, and 3D volumetric images. Each of these types of signals has its properties and advantages: dynamic 2D US images (i.e., videos) allow the physician to analyse the temporal variation of an anatomical feature (e.g., the movement of a muscle, the volume of the ventricle). This variation can be generated by the shift of the probe or by the movement of the anatomical part. 3D volumetric images offer a complete understanding of the spatial anatomy of the acquired region compared to a more expensive tool for acquisition (e.g., automatic 3D probes). US 2D images are simple to acquire and process, but offer a limited view of the anatomical part of interest.

In this context, we discuss a real-time method [3] for denoising ultrasound 2D videos and 3D images. We extend the framework proposed for US 2D images, training an image-to-image learning network to replicate the behaviour of the denoising algorithms (e.g., BM4D, Vidosat). The network learns to predict images, which have been denoised through an algorithm that exploits the properties of the signal (e.g., temporal/spatial redundancy); then, the 2D videos or 3D images are reconstructed as ordered sequences of frames/slices. This approach applies a standard learning architecture, still exploiting the characteristics of the denoising algorithm while avoiding using more complex architectures (e.g., spatio-temporal networks) and managing videos of different lengths for the network's training.

S. Cammarasana and G. Patanè, *Real-Time Processing of Ultrasound Images*, Synthesis Lectures on Visual Computing: Computer Graphics, Animation, Computational Photography and Imaging, https://doi.org/10.1007/978-3-032-05805-8_5

The training phase can also be specialised to an anatomical district and input signals. The real-time computation depends only on the execution time of the network prediction. The method is general in terms of the input data (e.g., 3D images, time-dependent US videos), type of noise (e.g., speckle, Gaussian noise), resolution (e.g., isotropic, anisotropic), of the input signals, acquisition methodology, and anatomical district.

We present a real-time deep learning framework for denoising (Sect. 5.1) and the extension to 2D videos and 3D images (Sect. 5.2).

5.1 Real-Time Deep Learning Framework

We discuss a general deep-learning framework for the real-time denoising of ultrasound images (RT-DEN) (Fig. 5.1). The framework is general in terms of the input data, resolution of the input images (e.g., isotropic, anisotropic images), acquisition methodology, anatomical district, and speckle noise intensity. The framework is also general in terms of the building blocks and parameters of the deep learning framework; in fact, we can select different denoising algorithms (e.g., WNNM, SARBM3D) and deep learning architectures (e.g., Pix2Pix, VGG19). After designing a training data set composed of raw images and the corresponding denoised images, RT-DEN trains a neural network replicating the denoising results. Then,

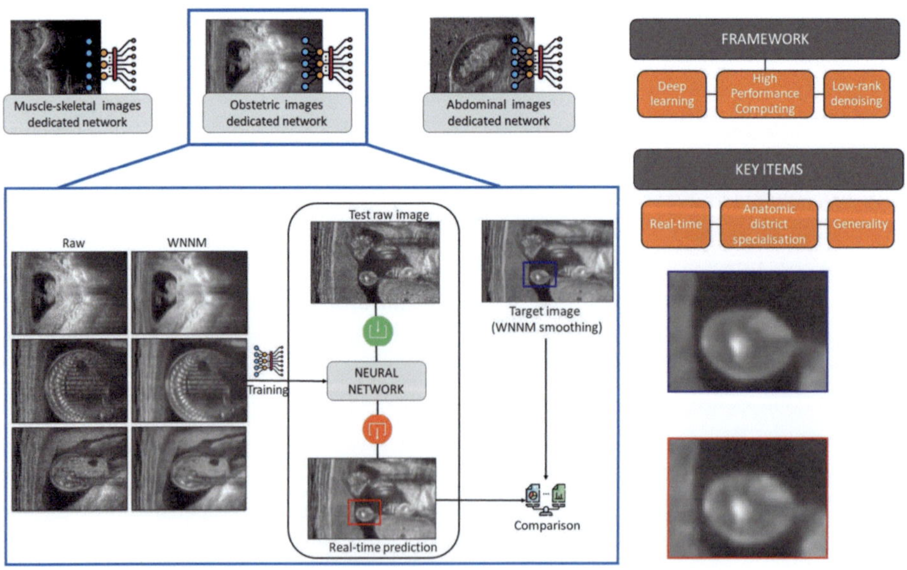

Fig. 5.1 Pipeline of the real-time denoising framework, with a magnification of prediction (red) and target (blue) images (right side). The framework applies deep learning and HPC to learn and replicate the denoising results of the state-of-the-art low-rank method (i.e., tuned-WNNM), in real-time and with a specialisation to anatomic districts. Image courtesy of [4]

real-time denoising is achieved through the prediction of the trained network. The framework combines three elements: *low-rank denoising*, *deep learning*, and *high-performance computing* (HPC). RT-DEN selects tuned-WNNM- as the best denoising method, according to the qualitative and quantitative comparison of five denoising methods, i.e., WNNM, SAR-BM3D—*SAR Block-Matching 3D* [17], BM-CNN—*Block Matching Convolutional Neural Network* [2], NCSR—*Non-Locally Centralised Sparse Representation* method [7], PCA-BM3D *Principal Component Analysis Block Matching 3D* [6]).

To achieve a *real-time denoising of ultrasound images*, the deep learning framework is based on the learning of the tuned-WNNM and HPC tools. The training is performed offline and can be further improved with new data, a priori information on the input images or the anatomical district, and denoised images selected after experts' validation. The execution time of denoising only depends on the network prediction, achieved in real time on standard medical hardware. As an experimental comparison, a quantitative (e.g., PSNR, SSIM) and qualitative evaluation is performed on the selected denoising methods on ultrasound images acquired from different anatomical (e.g., musculoskeletal, obstetric, and abdominal) districts. Then, the deep learning and HPC framework results are quantitatively and qualitatively analysed on a large collection of ultrasound images.

We introduce the real-time deep learning framework for 2D images denoising in terms of methodology (Sect. 5.1.1), learning architecture (Sect. 5.1.2), and experimental results (Sect. 5.1.3).

5.1.1 RT-DEN Method

The real-time denoising is based on training a neural network to learn and replicate the tuned-WNNM behaviour. LLR can also be inserted into the framework in place of *tuned-WNNM* for the real-time denoising of ultrasound 2D images. In the first phase, the network is trained on a dataset of ultrasound images of the same district. The data set for training the learning method comprises a set of pairs of ultrasound images: the input (i.e., the raw image) and the target (i.e., the image denoised with the tuned-WNNM filter). The ground truth is not available in ultrasound applications; for this reason, the target of the learning method is the output of the tuned-WNNM filter. Then, the trained network provides the denoised output through a real-time prediction of the test images. The deep learning framework is general in terms of the input data, i.e., type of noise (e.g., speckle, Gaussian noise), resolution (e.g., isotropic, anisotropic) of the input images, acquisition methodology, and the anatomical district. RT-DEN is also general in terms of building blocks and parameters. Since different methods (e.g., NCSR, SAR-BM3D, custom methods) have good performances, the generality of the framework allows the user to apply a different denoising algorithm and exploit its distinct characteristics in denoising and edge preservation. Alternative denoising methods can be used for various types of noise (e.g., speckle, Gaussian noise) and signals (e.g., 3D images, time-dependent ultrasound videos). The training phase is optimised for specific anatomical districts or types of noise. For instance, RT-DEN trains a specific network for each district, thus obtaining a more precise result when predicting the denoised image, as each network is specialised to the input anatomical features.

Deep Learning for Image-to-Image Regression

The *VGG19* [23] introduces *Convolutional Neural Networks* (CNN) pushing the depth to 16–19 weight layers, using small convolution filters of 3×3 size, with an application to large-scale image classification. The *Pix2Pix* [9] method is a *Generative Adversarial Network* (GAN), where the generator is a U-net [20], the discriminator is an encoding network [12], and the loss function is based on the binary cross-entropy. The *deep convolutional generative adversarial network* [18] applies unsupervised learning for image classification and the generation of natural images, exploiting batch normalisation, rectified linear unit activations, and removing fully connected hidden layers. CycleGan [25] learns to translate between domains without paired input-output examples. Adding a perceptual loss as an optimisation target of the generator [8] allows the network to improve structure preservation. Introducing a spectral normalisation as a weight normalisation technique [16] allows stabilising the discriminator's training in a GAN. For a further discussion of image-to-image deep learning methods, we refer the reader to [11].

5.1.2 Deep Learning Architecture

RT-DEN specialises Pix2Pix to ultrasound images, with two updates: (i) a validation data set of the same district of the training data set, which forces the exit condition when the validation error increases, and (ii) padding and masking pre-processing operations, which allow the framework to deal with images of different resolution, without an image resize that would imply a distortion artefact. The Pix2Pix architecture requires input images of constant size. Through padding, the images are resized according to the input needed for the architecture; after that, a masking filter allows the network to process only the numerical values of the image and not the values added by padding. A comparison between Pix2Pix and *Matlab* CNN is discussed in Sect. 5.1.3.

HPC Framework for Learning

RT-DEN defines an HPC implementation of the deep learning framework, taking advantage of a large ultrasound data set with 10K ultrasound images and of the CINECA-Marconi100 cluster, exploiting both CPUs (IBM POWER9 AC922) and GPUs (NVIDIA Volta V100). Given a training data set composed of raw images and the corresponding denoised images, RT-DEN implements a parallel and distributed deep learning framework in TensorFlow2.

Then, it defines a batch file for the execution of the deep learning framework on the cluster, which specifies the number of nodes, CPUs, GPUs, and memory of the cluster. Through the HPC framework, RT-DEN trains multiple networks with large data sets in a reasonable time for the target medical application, thus increasing the specialisation to anatomical districts and, consequently, the accuracy of the deep learning framework. The HPC framework generates a network model that is stored and used for predicting the output results. Furthermore, RT-DEN improves the offline training with new data, a priori and additional information on the input data (e.g., input anatomical district, noisy type/intensity, image resolution, acquisition methodology/protocol). The existing training data set update is always performed offline by adding new images after expert validation of the denoising results.

Training Data Sets and Metrics

We consider the US data set, which contains more than 10K ultrasound images at different resolutions acquired from different anatomical districts (e.g., musculoskeletal, obstetric, abdominal). In this data set, we compare the performance of varying denoising methods applied to ultrasound images and analyse the framework's performance through the training and the prediction of the learning-based network, with a specialisation to anatomic districts. We generate and test different data sets by varying the number of images for the training phase and the anatomical district for the prediction phase. The custom Pix2Pix network is trained on four data sets of obstetric images: (a) 500, (b) 1500, (c) 3500, and (d) 5000 images. Each data set comprises the input images (i.e., the raw US data set) and the target images (i.e., the corresponding images denoised with the tuned-WNNM). The validation data set comprises an additional set of different images (i.e., about 10% of the training data set) of the same district. Then, we evaluate each of the four networks (i.e., the networks trained with a different number of images) with two other test data sets of 50 images each, respectively, from the (i) obstetric and (ii) muscular anatomical districts.

5.1.3 Experimental Results

We present the prediction results of the real-time denoising through a quantitative and qualitative evaluation, analysing the properties of the training dataset and the learning architecture and discussing the execution time and computational cost.

Qualitative Results

The predicted images are very close to the target image in all four cases; the network reproduces the edges and grey-scale values well. Furthermore, the predictions do not generate artefacts or spurious patterns. Varying the number of images of the training data set from

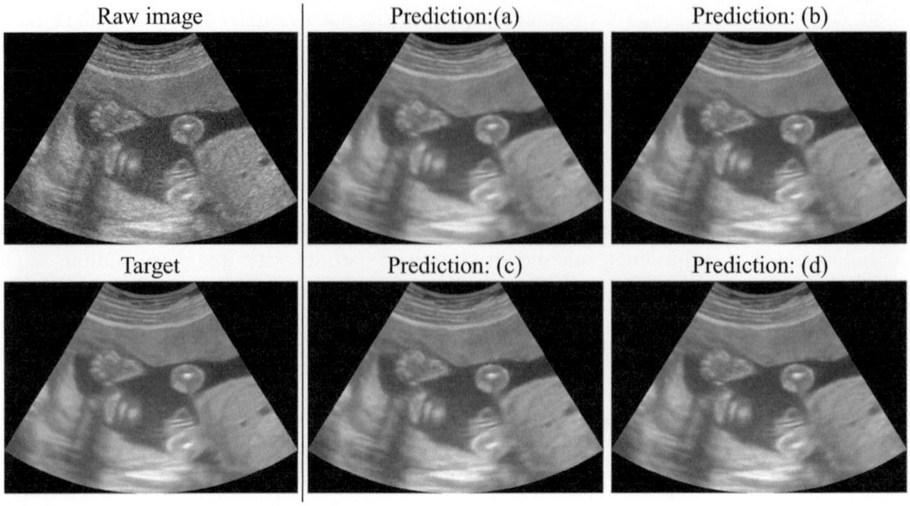

Fig. 5.2 Raw, target, and prediction images related to the obstetric data set (i). Training set: **a** 500 images, **b** 1500 images, **c** 3500 images, **d** 5000 images (Sect. 5.1.2). Image courtesy of [4]

500 to 5 K (Fig. 5.2a–d), the predicted images are slightly better than the target denoised images. Nevertheless, the results are good even with a small training data set of 500 images.

Figure 5.3 shows the prediction results of the four networks when tested with musculoskeletal images (ii). Predicting the output images with the networks trained with obstetric images (Fig. 5.3a–d), the results are slightly worse than the corresponding case in Fig. 5.2, even if the predicted images do not show any artefact of pattern repetition. Figure 5.4 shows the prediction results of the four networks when tested with TUS-REC data set, where the predicted images do not show any hallucination effect. The predictions of musculoskeletal and arm districts are performed with networks that are trained with images from a different (i.e., obstetric) district. This result confirms that each district requires a specific network, and a single network for all the districts gives lower-quality results.

Quantitative Results

Table 5.1 reports the quantitative metrics (Sect. 2.4) computed between the target and the predicted images. The network trained with 5K images (d) tested with obstetric images (i) has a median PSNR and SSIM value of 36.13 and 0.964, respectively, while the same network tested with musculoskeletal images (ii) has a median PSNR and SSIM value of 26.58 and 0.881. Both metrics have a very slight improvement when passing from a training set of 500 to a training set of 3500 images, confirming the results of the qualitative analysis. An additional increase of the training data set size to 5K images further improves the quantitative results for both the test data sets. Figure 5.5 shows the PSNR and SSIM metrics box plot

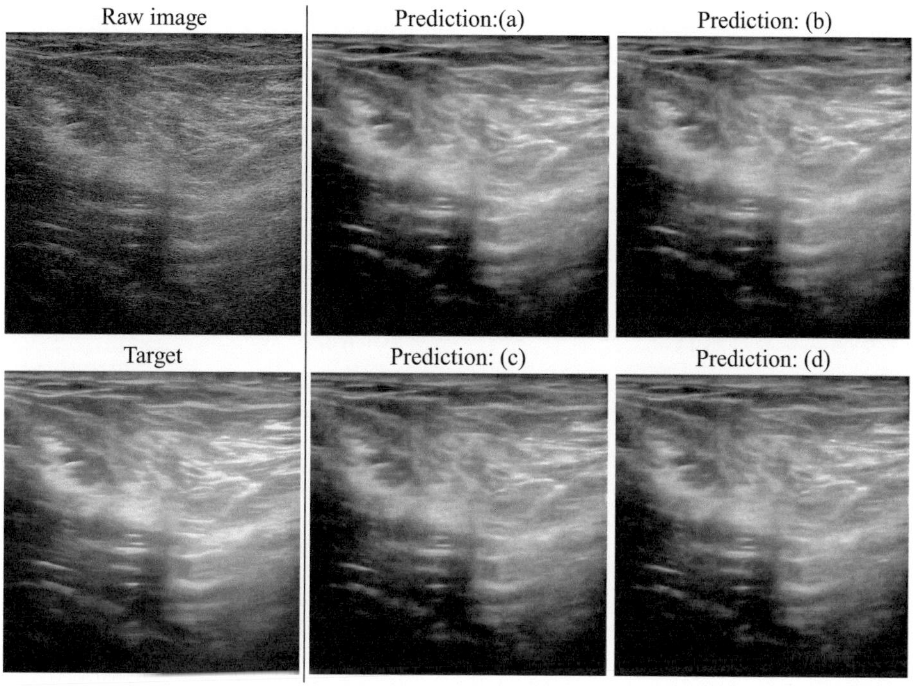

Fig. 5.3 Raw, target, and prediction images related to the musculoskeletal data set (ii). Training set (Sect. 5.1.2) with **a** 500, **b** 1500, **c** 3500, and **d** 5000 images. Image courtesy of [4]

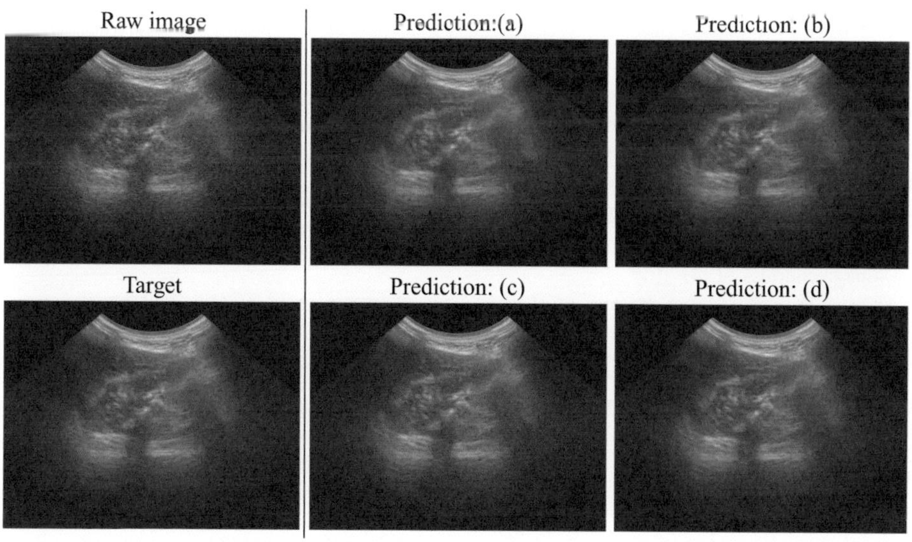

Fig. 5.4 Raw, target, and prediction images related to the arm data set (ii). Training set (Sect. 5.1.2) with **a** 500, **b** 1500, **c** 3500, and **d** 5000 images

Table 5.1 Concerning the four training data sets and the two test data sets (i.e., obstetric: *Ob.*, and musculoskeletal: *Msk.*) described in Sect. 5.1.2, we report the PSNR and SSIM metrics computed between the target and the prediction images as median values among the 50 test images

Metrics	PSNR		SSIM	
Test data set	Ob.	Msk.	Ob.	Msk.
Training with				
(a) 500 images	35.93	25.88	0.973	0.886
(b) 1500 images	34.52	26.33	0.957	0.854
(c) 3500 images	36.07	26.31	0.962	0.878
(d) 5000 images	36.13	26.58	0.964	0.881

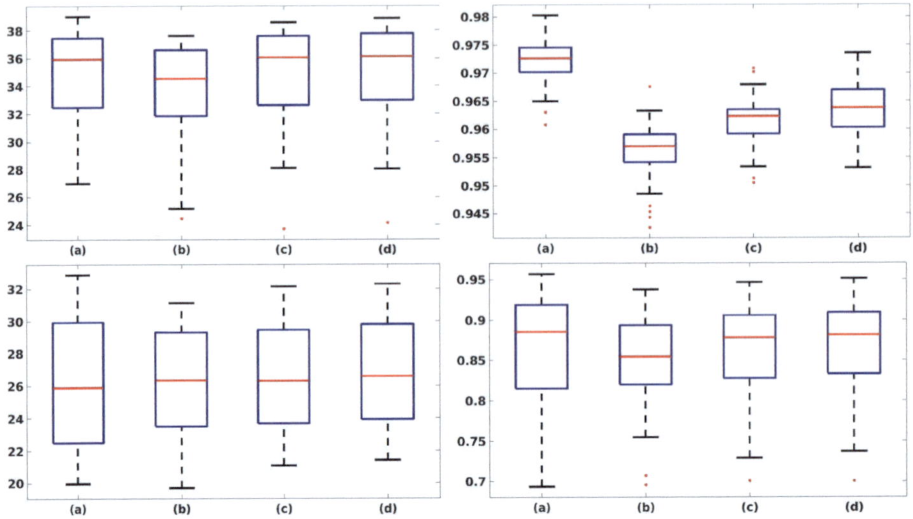

Fig. 5.5 PSNR and SSIM boxplot for each of the four training data set (i.e., (**a**–**d**)) and the two test data sets (Sect. 5.1.2): (top-left) PSNR, obstetric test data set; (top-right) SSIM, obstetric test data set; (bottom-left) PSNR, musculoskeletal test data set; (bottom-right) SSIM, musculoskeletal test data set. Image courtesy of [4]

for four training and two test data sets. Increasing the number of images of the training data set, the range of the metrics tends to decrease; this behaviour has a lower variability on the prediction of the output image. These results confirm that a network specialised in a single anatomical district reaches the best denoising quality. The prediction of musculoskeletal images from a network trained with obstetric images highly reduces the performance of the real-time denoising framework; in fact, the network learns that it replicates not only the denoising algorithm itself but also its adaptation to the anatomic structures and features of each district.

Single Versus Multiple Districts

We compare the prediction results of the real-time denoising framework with three different training data sets. The first two data sets have 500 and 1500 ultrasound images of the same district (i.e., the obstetric one). The third one comprises 1500 images of different districts; in particular, we select 500 images from the cardiac, obstetric, and musculoskeletal districts. Due to the various resolutions of the images, padding has been applied to obtain the same input resolution for each network. We evaluate the prediction results on four test data sets: the first three are composed of 50 images from the cardiac, obstetric, and musculoskeletal districts, respectively. The fourth comprises 50 images randomly selected from the three districts. The prediction results (Fig. 5.6 and Table 5.2) show that the networks trained with obstetric images (i.e., the single district networks) give the best results with the obstetric test data set: the predicted image of the single district network shows fewer scattering artefacts than the multiple districts network. Also, the single district networks have better results in terms of quantitative metrics: adding further images from different districts to the training data set worsens the results; in fact, the single district network with 500 obstetric images has a PSNR value of 35.93, while the multiple districts network has a PSNR value of 33.70.

Comparing the results on the other test data sets (e.g., cardiac Fig. 5.7 and Table 5.2), the network trained with images of multiple districts has better results than the networks trained with obstetric images only. The multiple district network better generalises on the

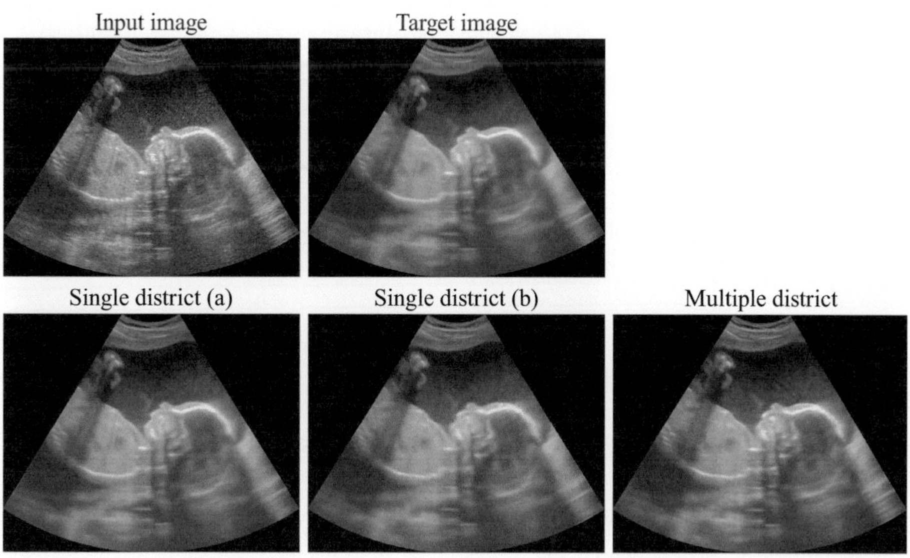

Fig. 5.6 Prediction results of the obstetric district, with the networks trained with 500 (**a**) and 1500 (**b**) images from the obstetric district and 1500 images from multiple districts (500 obstetric, 500 cardiac, 500 musculoskeletal images). Image courtesy of [4]

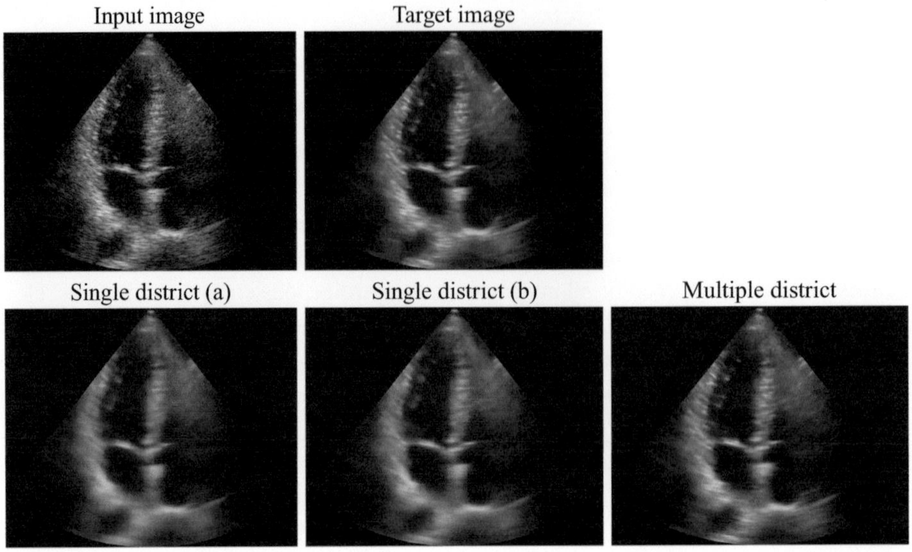

Fig. 5.7 Prediction results of the cardiac district, with the networks trained with 500 (**a**) and 1500 (**b**) images from the obstetric district and 1500 images from multiple districts. Image courtesy of [4]

Table 5.2 Concerning the results in Figs. 5.6 and 5.7, we report the PSNR metric computed between the target and the prediction images, as the average value among the 50 images of each test data set: obstetric (Ob.), musculoskeletal (Msk.), cardiac (Card.), and multiple districts (Multi.). The networks are trained with single district (a, 500 obstetric images), single district (b, 1500 obstetric images) and multiple district images

Test data set	Ob.	Msk.	Card.	Multi.
Training with				
Single district (a)	35.93	25.88	28.46	29.47
Single district (b)	34.52	26.33	28.63	29.51
Multiple district	33.70	28.41	33.82	30.33

denoising algorithm than on the anatomic features, thus generating fewer artefacts in the prediction. Furthermore, the multiple district network includes 500 cardiac images in its training dataset, improving the prediction results in this district. These features allow the multi-district network to reduce the artefacts on the edges of the anatomical features. A similar result is present in the musculoskeletal test data set, where the results of the network trained with multi-district images generalise better than the network trained with obstetric images; in fact, the single district network with 1500 obstetric images has a PSNR value of 26.33, while the multiple districts network has a PSNR value of 28.41. As the main conclusion, a dedicated network for each anatomic district is the best solution for predicting the denoised ultrasound images of each specific district if a sufficiently large data set is available for the training.

Comparison Between Different Deep Learning Architectures

As part of the deep learning framework, we compare the prediction results of two different networks: Pix2Pix and the (*Matlab*) CNN. Figure 5.8 shows that Pix2Pix has better results than CNN regarding blurring reduction, noise removal, and edge preservation. We also compare the quantitative metrics between the target images and the predicted images on the test data set of 50 obstetric anatomic district ultrasound images (Sect. 5.1.2). Pix2Pix has a PSNR average value of 36.07 and an SSIM average value of 0.878, while CNN has a PSNR average value of 25.69 and an SSIM average value of 0.651. This result underlines that Pix2Pix outperforms CNN as a network architecture for the deep learning framework.

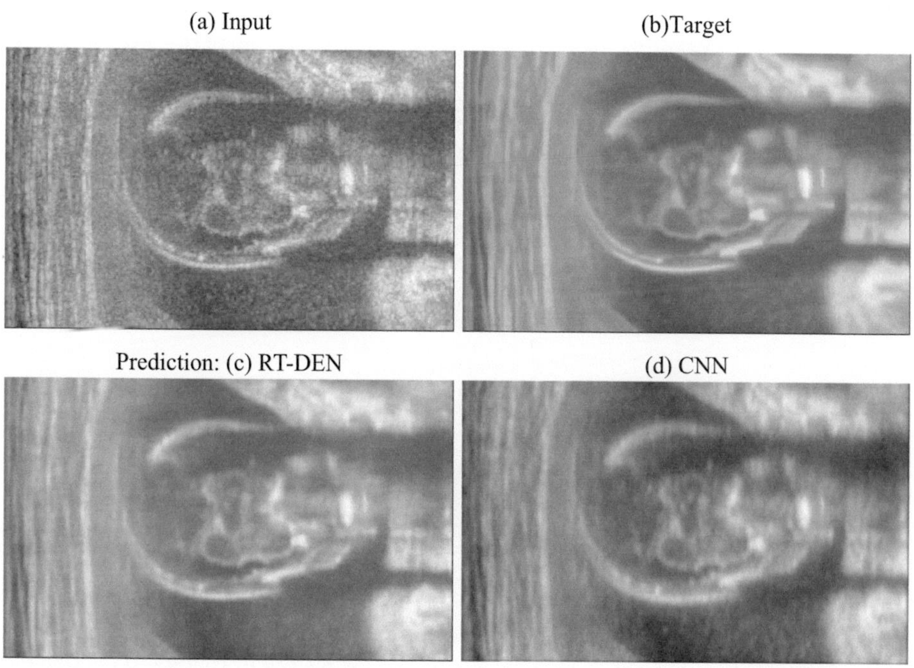

Fig. 5.8 **a** Input, **b** target, **c** RT-DEN prediction with Pix2Pix, and **d** prediction with CNN for the obstetric district. Image courtesy of [4]

Execution Time and Computational Cost

To test the training phase of the deep learning framework on the HPC environment, we exploit 8 nodes, each one composed of 32 cores and 4 accelerators, for a theoretical computational performance of 260 TFLOPS and 220 GB of memory per node. The parallel implementation of the deep learning framework and the high hardware performance reduce the computation time of the training phase by at least 100 orders less than a serial implementation on a standard workstation. The prediction's execution time is crucial for the framework's real-time implementation. We test the denoising prediction on GPU-based hardware, replicating the hardware of an ultrasound scanner currently in use. Given a set of ultrasound input images from different districts, the average execution time is 25 milliseconds; this result confirms that we achieved the real-time computation target required by the industrial constraint. After the industrialisation of the framework, the execution time can be further reduced with a hardware-oriented implementation that optimises the prediction operations of the neural network in terms of computation. We underline that the input resolution of the network is 600×600, which is reached through the zero-padding of each input image. The computational cost of the prediction depends on the resolution of the input image and the architecture of the network: in particular, the computational cost of a convolution operation is $O(r/s_r \cdot c/s_c) \cdot (f_r \cdot f_c) \cdot f$; in the application, the input image has a resolution of $r = c = 600$, the kernel-filter size on rows and columns is $f_r = f_c = 4$, the stride on rows and columns is $s_r = s_c = 2$, we use 10 convolution and 10 deconvolution operators, and several kernel-filters from 32 to 512.

5.2 2D US Video and 3D Image Denoising

The extension *Block Matching 4D* (BM4D) [13] of the BM3D algorithm to volumetric images exploits grouping and collaborative filters, where similar voxels are stacked together into 4D groups. A variant of non-local means denoising to 3D biomedical images [15] considers the spatially varying noise levels across the volumetric data. It automatically adjusts the strength of the filter through a noise estimation approach. The K-SVD algorithm [1] represents the signal as a sparse linear combination of atoms from an over-complete dictionary; the atoms are iteratively updated through an SVD of the representation error to fit the data better. A 3D blockwise version of the non-local means filter with wavelet sub-bands [5] is applied to biomedical images. A learning-based method is used to denoise 3D biomedical data [19] through a residual encoder-decoder Wasserstein generative adversarial network.

Vidosat [24] applies online transform learning to 2D video denoising; the patches are computed either from corresponding 2D patches in successive frames or using an online block matching technique, fully exploiting the spatio-temporal data correlation. In VBM3D [14], 3D spatio-temporal volumes are constructed by tracking blocks along trajectories defined by the motion vectors. Then, the denoising of the blocks is achieved through collaborative filtering by transforming each block through a decorrelating 4D separable transform; finally, the shrinkage and inverse transformation generate the denoised output. The formulation of the denoising of mixed noise as a low-rank matrix completion problem [10] leads to a denoising scheme without strong assumptions on the statistical properties of the noise. Applying the non-separable oriented 3D dual-tree wavelet transformation [22] gives a motion-based multi-scale decomposition for video and isolates in its sub-bands motion along with different directions.

We introduce the real-time framework for 2D videos and 3D images denoising (Sect. 5.2.1) and the experimental results (Sect. 5.2.2).

5.2.1 Real-Time Framework for 2D Videos and 3D Images

To achieve real-time computation and maintain good denoising and edge preservation results, the deep learning and HPC framework for 2D images is extended to 3D images and 2D videos. In this application, the framework learns and replicates the dedicated denoising algorithms with an image-to-image prediction; this allows the user to apply a denoising method that accounts for the specific signal's properties with real-time execution. The framework presented in Sect. 5.1 trains a neural network to learn and replicate the behaviour of the denoising algorithms applied to arbitrary signals. In the first phase, different input signals are denoised through dedicated algorithms (e.g., BM4D, Vidosat), which exploit the spatio-temporal redundancy. Then, the network is trained on a collection of pairs of US images: the input (i.e., the raw image) and the target (i.e., the image denoised with the proper filter). The framework trains a specific network for each district, thus obtaining a more precise result when predicting the denoised image, as each network is specialised to the input anatomical features. Furthermore, images from different input signals (e.g., 2D videos and 3D images) are used to train specific networks. The real-time denoising depends only on the execution time of the network prediction.

WNNM is natively applicable to 2D images, and it is applied to 2D videos through frame-by-frame denoising and to 3D volumetric images through slice-by-slice denoising. Applying WNNM to 2D videos/3D images does not exploit any knowledge of the temporal/spatial redundancy of the signal. BM4D is natively applicable to 3D volumetric images, which will be applied to 2D videos; in this case, the temporal dimension is assumed as the third dimension of the volumetric data. Two main approaches are applied; given a 2D video $m \times n \times t$, where $m \times n$ is the resolution of a single frame, and t is the number of frames, BM4D can be applied to

- the full video: in this case, the $m \times n \times t$ tensor is the input video, and the output is a tensor representing the smoothed video;
- consecutive portions of the video: in this case, the smoothing of each frame t_i is performed through the denoising of the tensor $m \times n \times 2k$, where 2k is the interval of frames centred

in i, from $i-k$ to $i+k$. This procedure is repeated starting from the frame k for each video frame, i.e., $t-k$ times.

Vidosat is natively applicable to 2D videos but can be applied to volumetric images by considering the third spatial dimension as the temporal dimension of dynamic data. Both full video and consecutive portions (i.e., blocks of frames) can be considered input to the denoising algorithm.

5.2.2 Experimental Results

We present the results of US 2D videos and 3D image denoising through dedicated algorithms and the prediction results of the real-time denoising framework. Figure 5.9 shows the denoising results of WNNM, BM4D, and Vidosat, with *US volumetric images*. Analysing the images sectioned in the slice direction, WNNM and BM4D have better noise removal and anatomical feature preservation results, while Vidosat has a slight scattering effect. To integrate this evaluation, we apply a ray-casting algorithm [21] to visualise a 2D rendering of the volumetric data (Fig. 5.10). This visualisation allows us to better analyse the consistency among slices; in fact, BM4D has a smoother rendering than WNNM, which

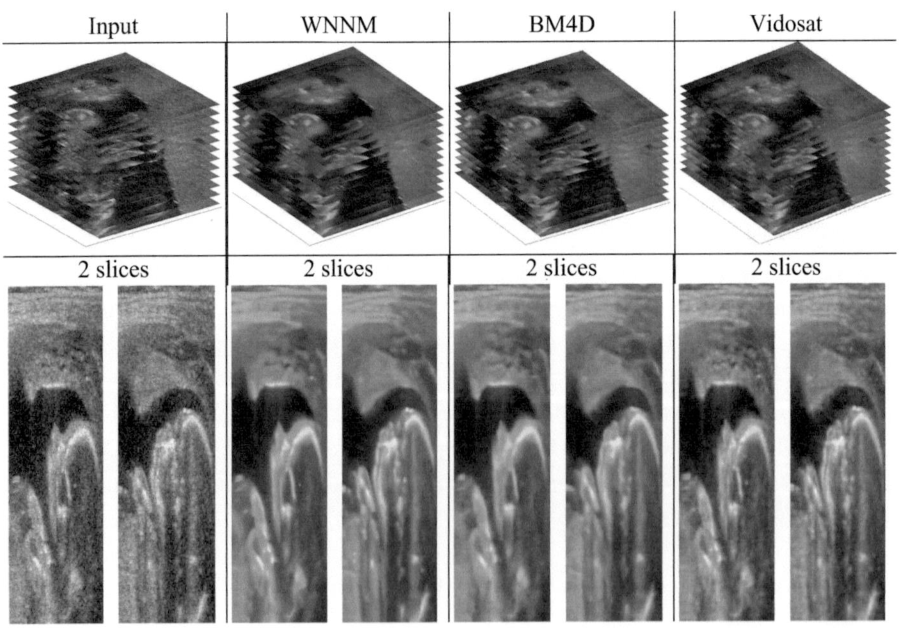

Fig. 5.9 Denoising of US volumetric image $312 \times 146 \times 36$ of the obstetric district. Image courtesy of [3]

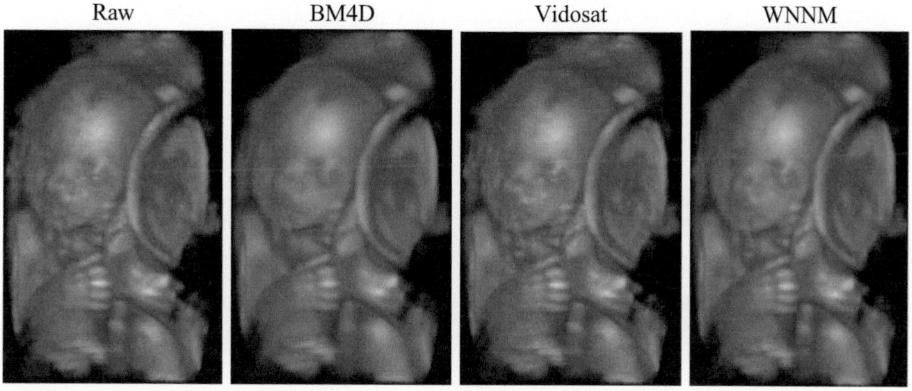

Fig. 5.10 Ray-casting [21] rendering of Fig. 5.9. Image courtesy of [3]

is affected by a scattering effect due to the phase shift between adjacent slices. Finally, Vidosat rendering is affected by the lower noise removal, and BM4D has the best results in denoising 3D volumetric images in terms of noise removal, anatomical feature preservation, and slice consistency. Figures 5.11, 5.12, and 5.13 show the denoising results on *2D videos*, on different anatomical districts: abdominal, cardiac, and musculoskeletal. We select the window-moving version of BM4D, WNNM is applied *frame by frame*, and Vidosat is used on the 2D video. Vidosat shows the best results regarding noise removal and anatomical feature preservation; the blurring effect is barely present. Since it is applied frame by frame, WNNM is unaffected by blurring effects. However, the noise reduction of WNNM is lower than BM4D and Vidosat since WNNM does not exploit any knowledge among consecutive frames. BM4D has a higher blurring effect and a lower noise reduction than Vidosat. Indeed, we selected BM4D and Vidosat to generate the target images from 3D images and 2D videos, respectively.

Qualitative Results

The learning-based framework allows the user to predict the denoised output of a 2D video (Fig. 5.14) and a 3D image (Figs. 5.15 and 5.16). In both cases, the predicted image is very similar to the target image; the output appears smooth, preserving anatomic features and edges. The predicted images do not present any artefacts or blurring effects. The smoothing intensity depends on the target images in the training dataset; images with higher smoothing generate a network that learns higher denoising of the input image. Furthermore, the framework allows the user to train several networks with different smoothing levels and select the preferred output among different choices.

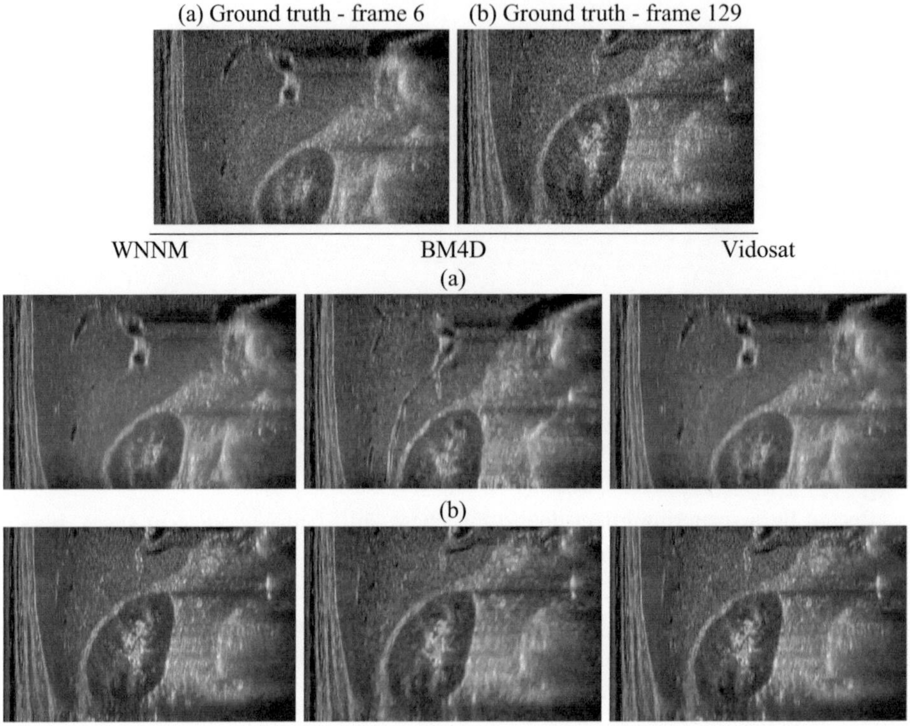

Fig. 5.11 Denoising of 2 frames of a 2D US video (560 × 359 × 256) of the abdomen. Image courtesy of [3]

Quantitative Error

Table 5.3 summarises the quantitative error computed between the target and the predicted images for different signals (e.g., 2D videos, 3D images) and anatomical districts (e.g., obstetric, cardiac) and measured through the PSNR and SSIM metrics. Images extracted from volumetric data and 2D videos are used to train separate networks. Each network is specific for an anatomical district (e.g., obstetric, cardiac). The selection of the anatomical regions depends on different parameters: the type of probe, the acquisition modality, the quantitative/qualitative analysis (e.g., segmentation, morphological evaluation), and the post-processing algorithms. Furthermore, different anatomical districts lead to varying types of signals: for example, in the cardiac district, the movement of the cardiac valve is acquired by the probe, while in the abdominal district, the movement of the probe is performed by the physician. The experimental results show that the framework can correctly denoise the signal through an image-to-image prediction for both 3D images and 2D videos. For example, the prediction of obstetric 3D images shows an average PSNR value of 33.15, while the prediction of cardiac videos shows an average PSNR value of 34.62.

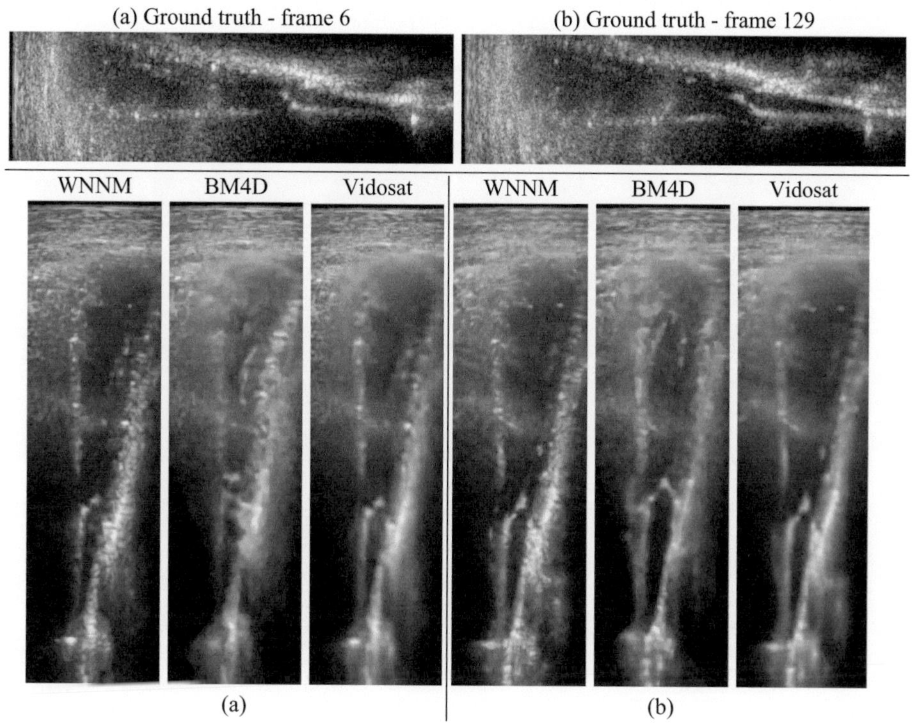

Fig. 5.12 Denoising of 2 frames of a 2D US video ($592 \times 168 \times 233$) of the cardiac district. Image courtesy of [3]

Execution Time and Computational Cost

Table 5.4 shows the *execution time* of the denoising methods with the different data sets (i.e., 2D video, volumetric image). BM4D is the fastest method, taking 75 seconds to process a $408 \times 120 \times 36$ volumetric image, while Vidosat and WNNM take 161 and 1260 seconds, respectively. None of these methods reaches the real-time requirement for denoising. The analysis of the execution time of the framework is divided into an *offline training* and an *online prediction*. To test the training phase of the deep learning framework on the HPC implementation, 8 nodes are exploited, each one composed of 32 cores and 4 accelerators, for a theoretical computational performance of 260 TFLOPS and 220 GB of memory per node. The parallel implementation of the deep learning framework reduces the computation time of the training phase by at least 100 orders of magnitude compared to a serial implementation on a standard workstation, with an average execution time of fewer than 0.1 seconds per epoch on a single image, i.e., about 2 hours on a data set of 500 images, with 150 epochs. We use 15 3D images of 36 stacks for the volumetric data set, for 540 2D images. For the video data set, we use several videos of different lengths for each anatomical district; we train each

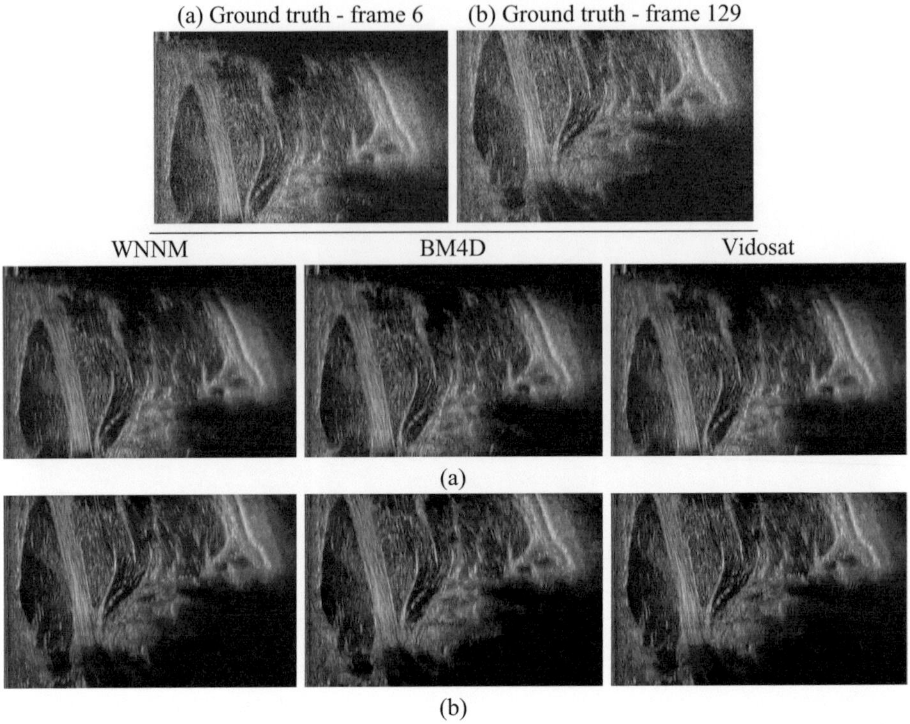

Fig. 5.13 Denoising of 2 frames of a US video (600 × 380 × 155) of the musculoskeletal district. Image courtesy of [3]

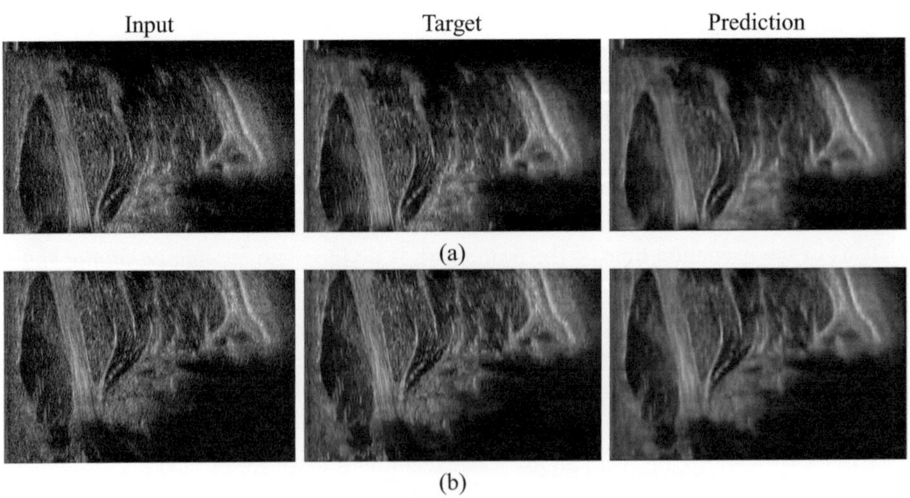

Fig. 5.14 Framework prediction for 2 musculoskeletal frames (**a**, **b**) of a 2D video. Image courtesy of [3]

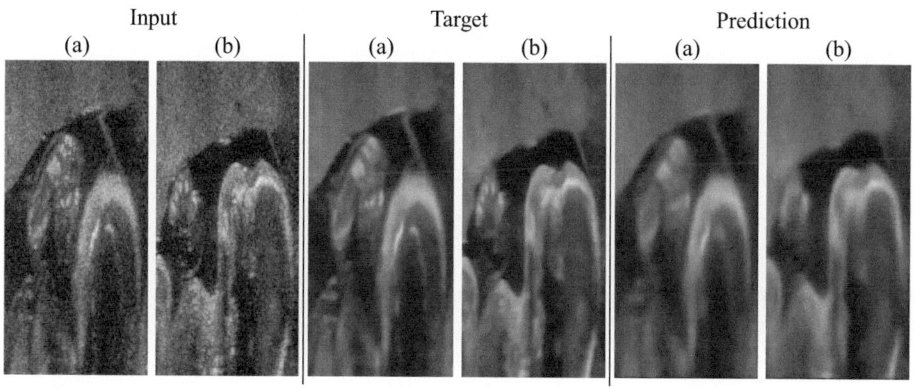

Fig. 5.15 Framework prediction for two slices (**a**, **b**) of a 3D image of the obstetric district. Image courtesy of [3]

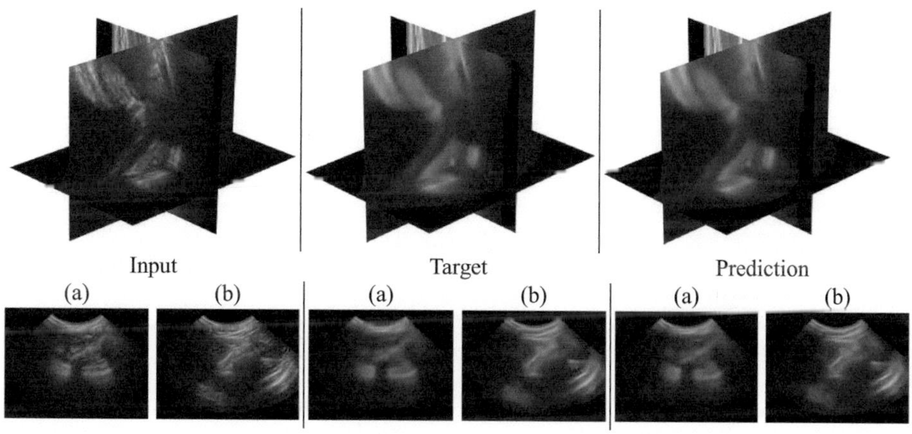

Fig. 5.16 Framework prediction for two slices (**a**, **b**) of a 3D image of the arm district

network with a minimum of 500 2D images. The execution time of the network prediction of the framework for the 2D image denoising is about 25 milliseconds, as already discussed in Sect. 5.1.3 and can be further optimised, thus confirming the real-time computation. The framework allows the user to denoise each image while acquiring the video/volumetric signal, since the spatio-temporal information is not required.

Table 5.3 PSNR and SSIM metrics computed between the target and the prediction image, as an average value among the 20 signals of each test data set: musculoskeletal videos (Msk.), cardiac videos (Card.), obstetric 3D images (Ob.)

Signal	Msk.	Card.	Ob.
PSNR	28.12	34.62	33.15
SSIM	0.963	0.978	0.970

Table 5.4 Concerning the tests in Figs. 5.9 and 5.13, we report the execution time of the denoising methods tested on 2D video and volumetric data

Method	WNNM	BM4D	Vidosat
2D video [s]	26820	1506	3338
3D image [s]	1260	75	161

References

1. M. Aharon, M. Elad, A. Bruckstein, K-SVD: an algorithm for designing overcomplete dictionaries for sparse representation. IEEE Trans. Signal Proc. **54**(11), 4311–4322 (2006)
2. B. Ahn, N.I. Cho, Block-matching convolutional neural network for image denoising (2017). arXiv:1704.00524
3. S. Cammarasana, P. Nicolardi, G. Patané, Fast learning framework for denoising of ultrasound 2D videos and 3D images, in *International Conference on Image Analysis and Processing* (Springer, 2022), pp. 475–486
4. S. Cammarasana, P. Nicolardi, G. Patanè, Real-time denoising of ultrasound images based on deep learning. Med. & Biolog. Eng. & Comput. 1–16 (2022)
5. P. Coupé et al., 3D wavelet subbands mixing for image denoising. Int. J. Biomed. Imag. **2008** (2008)
6. K. Dabov et al., BM3D image denoising with shape-adaptive principal component analysis, in *Signal Processing with Adaptive Sparse Structured Representations*, ed. by R. Gribonval (2009)
7. W. Dong et al., Nonlocally centralized sparse representation for image restoration. IEEE Trans. Image Proc. **22**(4), 1620–1630 (2012)
8. L.A. Gatys, A.S. Ecker, M. Bethge, Image style transfer using convolutional neural networks, in *Proceedings of the IEEE Conference on Computer Vision and Pattern Recognition* (2016), pp. 2414–2423
9. P. Isola et al., Image-to-image translation with conditional adversarial networks, in *Proceedings of the IEEE Conference on Computer Vision and Pattern Recognition* (2017), pp. 1125–1134
10. H. Ji et al., Robust video denoising using low rank matrix completion, in *Computer Society Conference on Computer Vision and Pattern Recognition* (IEEE, 2010), pp. 1791–1798
11. S. Kaji, S. Kida, Overview of image-to-image translation by use of deep neural networks: denoising, super-resolution, modality conversion, and reconstruction in medical imaging. Radiolog. Phys. Technol. **12**(3), 235–248 (2019)

12. A. Krizhevsky, I. Sutskever, G.E. Hinton, Imagenet classification with deep convolutional neural networks. Commun. ACM **60**(6), 84–90 (2017)
13. M. Maggioni et al., Nonlocal transform-domain filter for volumetric data denoising and reconstruction. IEEE Trans. Image Proc. **22**(1), 119–133 (2012)
14. M. Maggioni et al., Video denoising, deblocking, and enhancement through separable 4-D nonlocal spatiotemporal transforms. IEEE Trans. Image Proc. **21**(9), 3952–3966 (2012)
15. J.V. Manjón et al., Adaptive non-local means denoising of MR images with spatially varying noise levels. J. Magn. Reson. Imag. **31**(1), 192–203 (2010)
16. T. Miyato et al., Spectral normalization for generative adversarial networks (2018). arXiv:1802.05957
17. S. Parrilli et al., A nonlocal SAR image denoising algorithm based on LLMMSE wavelet shrinkage. IEEE Trans. Geosci. Remote Sens. **50**(2), 606–616 (2011)
18. A. Radford, L. Metz, S. Chintala, Unsupervised representation learning with deep convolutional generative adversarial networks (2015). arXiv:1511.06434
19. M. Ran et al., Denoising of 3D magnetic resonance images using a residual encoder-decoder Wasserstein generative adversarial network. Med. Image Anal. **55**, 165–180 (2019)
20. O. Ronneberger, P. Fischer, T. Brox, U-Net: convolutional networks for biomedical image segmentation, in *International Conference on Medical Image Computing and Computer-Assisted Intervention* (Springer, 2015), pp. 234–241
21. S.D. Roth, Ray casting for modeling solids. Comp. Graph. Image Proc. **18**(2), 109–144 (1982)
22. I.W. Selesnick, K.Y. Li., Video denoising using 2D and 3D dual-tree complex wavelet transforms, in *Wavelets: Applications in Signal and Image Processing X*, vol. 5207 (International Society for Optics and Photonics, 2003), pp. 607–618
23. K. Simonyan, A. Zisserman, Very deep convolutional networks for large-scale image recognition (2014). arXiv:1409.1556
24. B. Wen, S. Ravishankar, Y. Bresler, VIDOSAT: high-dimensional sparsifying transform learning for online video denoising. IEEE Trans. Image Proc. **28**(4), 1691–1704 (2018)
25. J.-Y. Zhu et al., Unpaired image-to-image translation using cycle-consistent adversarial networks, in *Proceedings of the IEEE International Conference on Computer Vision* (2017), pp. 2223–2232

6 US Super-Resolution

The spatial resolution of a US system is limited by the diffraction limit to length scales of approximately half of the wavelength of the transmitted beam; two or more points' reflectors that are spaced more closely than this limit cannot be distinguished. For the real-time US visualisation and enhancement based on super-resolution, it is fundamental to address the US image resolution that is affected by the required frequency of the video, since some anatomical districts (e.g., cardiac) need a high acquisition frequency to accurately acquire the behaviour of anatomical features that quickly change over time. The US image super-resolution is relevant to improving the quality of the image, its visual interpretation by the physician, and post-processing steps, e.g., classification, diagnosis, and segmentation. We discuss the real-time super-resolution of US signals, focusing on super-resolution for reconstructing the non-acquired scan lines. We perform a quantitative and qualitative comparison of super-resolution methods on an extensive collection of US signals (i.e., 2D/3D images, 2D videos) acquired from different anatomical districts. In this context, the required characteristics of the super-resolution method are

- *features preservation*, including the edges and geometries of organs and tissues;
- *artefacts reduction* in terms of blurring and shading;
- *real-time* processing, due to the industrial and clinical requirements of US image processing methods;
- *up-sampling factors*, the super-resolution methods apply to lateral resolution only, with up-sampling factors of 2X and 4X;
- *anatomical districts*, with specialised methods for different anatomical organs, e.g., cardiac, musculoskeletal; and
- *input signal*, 2D and 3D images, 2D videos.

The key objective of image super-resolution is to address the limitations of imaging systems, mainly due to hardware capabilities and requirements for clinical processing of medical

S. Cammarasana and G. Patanè, *Real-Time Processing of Ultrasound Images*, Synthesis Lectures on Visual Computing: Computer Graphics, Animation, Computational Photography and Imaging, https://doi.org/10.1007/978-3-032-05805-8_6

images, such as descriptions of small anatomical landmarks or changes in the microvasculature around a tumour. SR is an ill-posed problem that extrapolates the HR output from the corresponding LR input. Furthermore, there are two additional limiting factors: (i) matrix arrays are costly and present some technological hurdles at higher frequencies, and (ii) the computational complexity of the SR algorithms rises as the scaling factor increases. Finally, the super-resolution of 2D images is generally limited by: (i) tissue motion is inherently 3D, and its correction necessitates sampling in all directions; (ii) complexity of the network and shape of 3D vessels, whose 2D section is generally low informative, (iii) the unfeasibility of reconstructing a volume through plane-by-plane imaging due to the long acquisition time for each plane and the potential higher user-dependency.

We discuss the image super-resolution in terms of main concepts (Sect. 6.1), related work (Sect. 6.2), its application to noisy images (Sect. 6.3), a learning-based method for the super-resolution of noisy images (Sect. 6.4), data sets and metrics (Sect. 6.5), and the experimental results (Sect. 6.6).

6.1 Image Super-Resolution: Main Concepts

The spatial resolution of an ultrasound imaging system is limited by the diffraction limit to length scales of approximately half of the wavelength of the transmitted beam; two or more point reflectors that are spaced more closely than this limit can not be distinguished. Through US videos, the physician analyses the temporal variation of an anatomical feature (e.g., the movement of a muscle, the volume of the ventricle), which can be generated either by the shift of the probe or by the movement of the anatomical part. 2D US videos are acquired through 2D probes, which capture sequences of images at a given frequency. The resolution of each image is affected by the required frequency of the video since some anatomical districts (e.g., cardiac) require a high acquisition frequency to accurately acquire the behaviour of anatomical features that quickly change over time. The underlying image has a resolution of $l \times d$, where l is the number of scan lines (i.e., the lateral resolution), and d is the depth of the acquisition of each scan line (i.e., the axial resolution). Axial resolution refers to the ability to discern two objects longitudinally adjacent to each other in the ultrasound image; lateral resolution refers to the ability to distinguish two objects adjacent to each other; the lateral resolution is usually lower than the axial resolution in ultrasound.

The resolution of the image in terms of lateral direction (i.e., the direction perpendicular to the US propagation along the beam line) is primarily determined by the width of the ultrasound beam and the number of elements (i.e., the piezoelectric crystals) that are activated to generate the US waves. Current probes vary the number of beamlines acquired by activating/deactivating piezoelectric crystals, thus reducing lateral resolution and image acquisition time. The axial resolution can be varied by changing the length and frequency of the pulses, affecting the ultrasound wave's penetration. In this context, we focus on lateral low-resolution image acquisition, reduce the acquisition time, and subsequently reconstruct the high-resolution image without losing information in terms of data depth. Finally, the US image super-resolution is relevant to improve the quality of the image, its visual interpretation by the physician, and post-processing steps, e.g., classification [2], diagnoisis [4], and segmentation [28].

6.2 Related Work

Deep Learning Super-Resolution of Generic Images

In recent years, deep learning methods for super-resolution have become widespread. A sparse representation of patch pairs over two dictionaries, one for both *low-resolution* (LR) and *high-resolution* (HR) images [43] captures the statistical dependencies between the sparsity patterns of the low-resolution coefficients of the corresponding representations. In [13, 14]. A fully convolutional neural network exploits a large filter size in the non-linear mapping layer. It uses three colour channels to train and predict HR images from LR images. A data clustering, which groups patches and a Bayes strategy for patch selection [49], provides a fast up-sampling based on external learning. The main novelties of the enhanced deep super-resolution network [27] are a simplification of the conventional residual network architectures and a multi-scale super-resolution network that reduces the model size. A *super-resolution generative adversarial network* (SRGAN) [26] applies a deep residual network with skip-connection and a perceptual loss between generated and target images. The reduction of artefacts of the previous method is addressed by the Enhanced SRGAN [62], which improves the network architecture, the adversarial and the perceptual loss, removes the batch normalisation layer, and applies the residual scaling and smaller initialisation values. The perceptual quality of ESRGAN is improved by the ESRGAN+ method [47] through a novel *Residual-in-Residual Dense Residual* block, which increases the network capacity without affecting its complexity. The introduction of weight normalisation and wider features before the rectified linear unit activation function [70] achieves good up-sampling results with a low computational cost. Additional methods, classified according to supervised/unsupervised approach and domain-specific applications, are discussed in [63].

Vision-Based Super-Resolution of Generic Images

The interpolating up-sampling with cubic kernels [20] offers high accuracy with low computational cost through appropriate boundary conditions and constraints on the kernel functions. The interpolated values are computed as the weighted average of pixels in the 2X2 (bilinear, [17]) or 4X4 (bi-cubic, [36]) neighbourhood. A fast implementation of bilinear and bi-cubic interpolations [21] is applied to *Field Programmable Gate Array* (FPGA), reducing the computational complexity and the FPGA resources while providing an excellent trade-off between image quality and calculation simplicity. After the training of two dictionaries for LR and HR patches [67], the similarity of the sparse representation of LR and HR patches with the respective dictionary is exploited to generate the high-resolution image. Anchored neighbourhood regression [55] and its improved version [54] learn a regression to correlate LR and HR images for each atom of the dictionary and precomputed neighbourhood. The search for recursive patches within an image [19] is extended by allowing geometric variations, incorporating the geometry by localising planes and accounting for plane parameters to estimate the deformation of recurring patches. Bivariate rational fractal interpolation [71] improves the approximation results with respect to polynomial kernels, preserving image edges and textures.

Learning-Based US Super-Resolution

The main novelties of the enhanced deep super-resolution network [27] are a simplification of the conventional residual network architectures and a multi-scale super-resolution network that reduces the model size. Exploiting the signal's sparsity in the Fourier domain, interpolating missing data [68] allows reconstructing the HR US image with a low computational cost. A *super-resolution generative adversarial network* (SRGAN) [26] applies a deep residual network with skip-connection and a perceptual loss between generated and target images. The reduction of artefacts of the previous method is addressed by the Enhanced SRGAN [62], which improves the network architecture, the adversarial and the perceptual loss, removes the batch normalisation layer, and applies the residual scaling and smaller initialisation values. The perceptual quality of ESRGAN is improved by the ESRGAN+ method [47] through a novel *Residual-in-Residual Dense Residual* block, which increases the network capacity without affecting its complexity. Applying the SRGAN to US images [7] preserves the anatomical structures and the speckle noise pattern, thus improving the perceptual quality of the upsampled images. Dilated convolution [34] extracts the internal recurrence information from the test image; this method upsamples LR images when LR-HR examples are reduced. Fully convolutional U-net [57] obtains high-resolution vascular images from high-density contrast-enhanced US signals. In [53], the deep learning method exploiting feature extraction blocks, repeating blocks, and upsampling layers applies an up-sampling factor in the range 2-8. A Self-supervised CycleGAN [32] only requires the LR US image and generates perceptually consistent up-sampling results. Combining CycleGAN, two-stage GAN, and the zero-shot super resolution [12], it is possible to obtain super-resolution images with low blurring artefacts.

In recent years, deep learning methods for super-resolution have become widespread. The *statistical prediction model* (SPM) applies a sparse representation of patch pairs over two dictionaries, one for both *low-resolution* (LR) and *high-resolution* (HR) images [43], and captures the statistical dependencies between the sparsity patterns of the low and high-resolution coefficients of the corresponding representations. In [13, 14], a fully convolutional neural network exploits a large filter size in the non-linear mapping layer. It applies three colour channels to train and predict HR images from LR images. A data clustering groups patches, and a Bayes strategy selects patches and provides a fast super-resolution based on external learning [49]. The main novelties of the *enhanced deep super-resolution* (EDSR) network [27] are a simplification of the conventional residual network architectures and a multi-scale super-resolution network that reduces the model size.

A *super-resolution generative adversarial network* (SRGAN) [26] applies a deep residual network with skip-connection and a perceptual loss between generated and target images. The reduction of artefacts of the previous method is addressed by the Enhanced SRGAN [61], which improves the network architecture, the adversarial and the perceptual loss, removes the batch normalisation layer, and applies the residual scaling and smaller initialisation values. The perceptual quality of ESRGAN is improved by the ESRGAN+ method [47] through a novel *Residual-in-Residual Dense Residual* block, which increases the network capacity without affecting its complexity. The introduction of weight normalisation and wider features before the rectified linear unit activation function [70] achieves good super-resolution results with a low computational cost. Additional methods, classified according to supervised/unsupervised approach and domain-specific applications, are discussed in [63].

Vision-Based US Super-Resolution

Learning-based methods suffer from artefacts and blurring when dealing with noisy signals. Several vision-based methods have been proposed over the years. The interpolating up-sampling with cubic kernels [20] offers high accuracy with low computational cost through appropriate boundary conditions and constraints on the kernel functions. In [3], a novel deconvolution-based method applies the maximum a posteriori estimation to the restoration of the tissue response and is validated with several tissue-mimicking phantoms with specific scatterer concentrations. The *Alternating Direction Method of Multipliers* [41] is applied to the super-resolution of US images, including deblurring and denoising [38] through a combination of ℓ_1 and ℓ_2 minimisation. In [69], a deconvolution method models the envelope radio-frequency and point spread function, which is robust to noise and does not require the knowledge of the centre frequency of the acquired signal, assuming a Gaussian distribution for both the unknown signal to be restored. In the point spread function [72], the reconstructed image is built through a posterior model with hybrid Gibbs sampling [16]. The properties of the decimation matrix in the Fourier domain [73] are exploited to solve the super-resolution problem with a ℓ_p-norm regulariser, with $p \in [1, 2]$. The envelope of the radio frequency signal [22] applies repetitive data in the non-local neighbourhood of samples.

The interpolating super-resolution with cubic kernels (i.e., *cubic convolution* [20], CC in short) offers high accuracy with a low computational cost through appropriate boundary conditions and constraints on the kernel functions. The interpolated values are computed as the weighted average of pixels in the 2×2 (bilinear, [17]) or 4×4 (bi-cubic, [36]) neighbourhood. A fast implementation of bilinear and bi-cubic interpolations [21] applies to *Field Programmable Gate Array* (FPGA), reducing the computational complexity and the FPGA resources while providing an excellent trade-off between image quality and calculation simplicity. After the training of two dictionaries for LR and HR patches [67], the similarity of the sparse representation of LR and HR patches with the respective dictionary is exploited to generate the high-resolution image. Anchored neighbourhood regression [55] and its improved version [54] learn a regression to correlate LR and HR images for each atom of the dictionary and precomputed neighbourhood. The search for recursive patches within an image [19] is extended by allowing geometric variations, incorporating the geometry by localising planes and accounting for plane parameters to estimate the deformation of recurring patches. Bivariate rational fractal interpolation [71] improves the approximation results with respect to polynomial kernels, preserving image edges and textures.

Device-Based US Super-Resolution

The second harmonic image [52] contains less noise and blur than the first harmonic image. Furthermore, the lateral resolution increases as the harmonic pulse is auto-focused because the higher harmonics are generated in the centre of the beam. Then, the image super-resolution is achieved by combining the first and second harmonic images. Both spatial and temporal deconvolutions [29] are achieved by accounting for the transmit and receive processes, electrical transducer characteristics, and transmit focusing laws. Combining phase-contrast imaging, angular spectral decomposition, and a super-resolution reconstruction technique [9], it is possible to recover the location and dimensions of objects smaller than the imaging wavelength. The reconstruction through generalised Tikhonov regularisation [25] is evaluated as a function of transmit-receive bandwidth and a focal number of the transducer by comparing the results with traditional B-mode imaging. The *Time-domain Optimised Nearfield Estimator* [59] assumes an observation model based on the superposition of spatial responses; then, a maximum a-posteriori estimation finds the distribution and amplitude of hypothetical targets that match the observed data with minimal target energy. As a further improvement, the *Diffuse Time-domain Optimised Near-field Estimator* [15] represents each hypothetical target in the system model as a diffuse region of targets rather than a single discrete target, thus inducing a better signal approximation. The cellular microscopy technique of multi-focal imaging[11] is applied to localise the unique position of the scatterer of the signal; three foci receive multiple overlapping curves; and a maximum likelihood estimation allows the identification of the source of the scatter.

Multi-frame US Super-Resolution

The *Bilinear Deformable Block Matching* [5], which is a registration method that accounts for the complex and deformable motion of soft tissues, is applied to reconstruct the HR image by exploiting the shifting property of the Fourier transform and the aliasing relationship between the continuous Fourier transform of the HR image and the discrete Fourier transform of LR images [39]. The use of deep learning for motion estimation among different frames [1] reduces the effect of noise and artefacts and reconstructs HR images from a sequence of LR images. The modelling of the spatial correlation of the speckle noise [10] is applied to standard reconstruction methods, with a tissue-mimicking phantom and co-registered multi-images.

Super-Resolution with Denoise Filter Methods

Image super-resolution and deblurring have been addressed together through a self-supervised approach that accounts for *scale equivariant imaging* (SEI) [50] and whose results are comparable with fully supervised learning. In [65], a convolutional transformer layer has been combined with a super-resolution and an edge-preserving feed-forward network, achieving an efficient, lightweight image super-resolution. In [33], a *multi-scale super-resolution* (MSSR) deep encoder and decoder network, which integrates a wavelet multi-resolution analysis, has been proposed for single image super-resolution and recovery of multi-scale edges. In [31], the image super-resolution has been combined with motion deblurring through a deep decoupled *cooperative learning and deblurring super-resolution* (CLDSR) model. Furthermore, convolutional neural networks and super-resolution have been combined to address pedestrian detection [42]. In [37], a recurrent neural network for image super-resolution is enriched with a *cross-scale non-local* (CSNL) attention module, which can find more cross-scale feature correlations within a single low-resolution image. In learning-based image super-resolution, the training dataset is generated by downscaling high-resolution images with a predetermined operation (e.g., bicubic). This approach cannot be applied to those scenarios where the super-resolution images are unavailable (e.g., low-resolution ultrasound images) or where the characteristics of the acquisition process (e.g., acquisition parameters) are unknown. In [35], an unpaired super-resolution method with noise correction/smoothing has been proposed based on a generative adversarial network oblivious to a data set of paired training images.

6.3 Super-Resolution of Noisy Images

The super-resolution of 2D images is primarily studied and widespread in many applications, such as biomedicine [24], astronomy [45], and industrial [46] contexts. In the literature, several methods guarantee excellent results in terms of reconstruction accuracy [51]. However, most of the current super-resolution methods do not account for noise in the image. At the same time, the preservation of the features and visual quality of the input data is affected by an underlying noise distribution. In contrast, these super-resolution methods generally smooth the noise with a blurring effect [40, 58] or apply a denoising filter before the super-resolution [18]. On the one hand, denoising may be required in image processing; on the other hand, noise reduction alters the visual quality of the super-resolution image. The processing of digital images affected by different types of noise is relevant in many application fields: biomedical (e.g., ultrasound imaging) [64], geo-[60] and astro-[23] physics, biology [56], computer vision [30], and art design [66]. The application of filters on noisy images without altering the noise component is relevant for preserving the information, maintaining input image quality and integrity, and increasing the robustness of the post-processing algorithms.

In this context, *noisy super-resolution* method (NSR) generates a high-resolution (HR) image from a low-resolution (LR) noisy image, preserving the visual quality and the quantitative similarity of the input data without altering the noise distribution. The design of a proper super-resolution method that accounts for noise allows us to improve the visual quality and reduce the artefacts (e.g., blurring and scattering) of the high-resolution predicted image. To this end, NSR defines a novel super-resolution of noisy images based on a learning model that accounts for the presence of the noise and preserves the similarity of the high-resolution image in terms of visual quality and quantitative metrics (e.g., normalised crossed correlation, normalised mean squared error, peak-signal-to-noise-ratio, structural similarity feature-based similarity, universal image quality). In contrast, previous work does not account for the noise distribution to preserve the similarity with the high-resolution image. For the learning of the model, NSR defines a data set of ground-truth, noisy, and down-sampled noisy synthetic 2D images, trains the network to up-sample the low-resolution noisy image, and matches the high-resolution noisy image. At the same time, the prediction has to preserve the noise properties with respect to the ground-truth image. NSR specialises the networks to different up-sampling factors, i.e., 2X and 4X, and noise, i.e., speckle and Gaussian. The approach is general to specialise with additional noise types (e.g., Poisson) and up-sampling factors (e.g., 8X).

The NSR method improves the super-resolution of noisy images with respect to state-of-the-art standard and learning-based methods. We evaluate several quantitative measures for local, structural, and quality similarity. For example, NSR achieves a PSNR average value of 20.64 with a 4X up-sampling factor and Gaussian noise, while previous works achieve an average 20.11 value; also, NSR gains an average mean-squares error value of 247 with a 2X up-sampling factor and speckle noise, while state-of-the-art methods achieve more than 400 as an average value. We evaluate the preservation of image properties, such as brightness, contours, and textures. NSR has results comparable to the standard method and improves learning-based methods on ultrasound images affected by speckle noise. We also analyse the distribution of the generated noise with respect to the ground-truth image; NSR generally better preserves the noise distribution. Trained networks for 2X and 4X up-sampling factors, Gaussian and speckle noise, are available at https://github.com/cammarasana123/noise-SuperResolution.

6.4 Learning-Based and Quality-Preserving Image Super-Resolution

WDSR [70] is a learning architecture that exploits residual blocks as it improves the prediction of images where the difference between the input and the target is small. NSR proposes a customised version of this network: *custom-WDSR*. In particular, the network architecture is a variant of WDSR-A, where the expansion of the features before the rectified linear unit (ReLU) activation allows more information to pass through while preserving the non-linearity of the network. After normalising the data, NSR applies a 2D convolution and a weighted normalisation that improves the conditioning of the optimisation problem and, thus, the convergence. Then, NSR applies eight residual blocks with wide activation, where each comprises two convolution layers with ReLU activation and a final 2D convolution with a weighted normalisation layer. Finally, NSR combines residual blocks and convolution layers, applies deconvolution to interpolate the missing lines and columns, and denormalisation to match the target image. The kernel filter size depends on the up-sampling factor: (3×3) in 2X up-sampling, and (5×5) in 4X up-sampling. With this setting, the total number of trained parameters is 889K for the 2X network and 253K for the 4X network (Fig. 6.1).

The network input is a noisy, low-resolution image, while the target is the corresponding high-resolution image. The loss function accounts for both the loss between the predicted image and the target image (i.e., the noisy image) and the log-likelihood of the generated noise with respect to the input noise. In particular, we define the prediction (**P**), noisy (**N**) and ground-truth (**G**) images, and the loss as

$$\mathcal{L} = \|\mathbf{P} - \mathbf{N}\|_F + \lambda \frac{1}{m} \sum \log(L(\mathbf{P} - \mathbf{G}, \Theta)), \tag{6.1}$$

where Θ is the known noise distribution (e.g., Gaussian) and m is the number of pixels of the high-resolution image. The first term (i.e., $\|\mathbf{P} - \mathbf{N}\|_F$) trains the network to match the target high-resolution noisy image, while the second term trains the network to generate a prediction whose noise distribution with respect to the ground-truth image matches the Θ

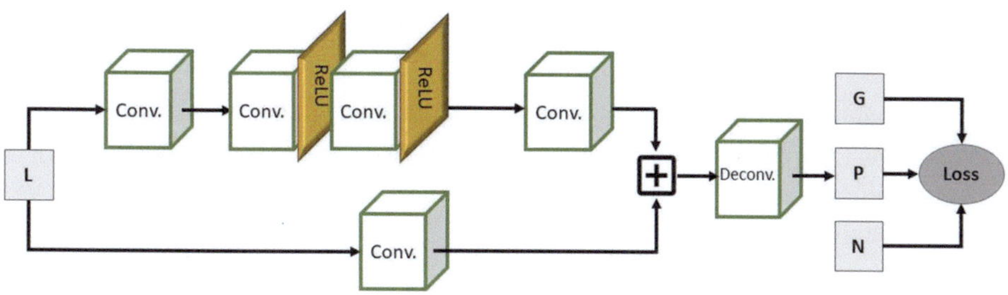

Fig. 6.1 Network architecture: noisy low-resolution (**L**), ground-truth (**G**), noisy high-resolution (**N**), prediction (**P**). Image courtesy of [6]

distribution; the second term is computed with the *log-loss* function L, and it is weighted through the parameter $\lambda = -10$. The λ term is negative as we maximise the noise distribution properties with respect to the Θ distribution. As additional parameters for the training, NSR applies the Adam optimiser with a learning rate of 0.001, a maximum number of epochs of 60, and an early stopping criterion by monitoring the validation loss. The networks are trained with 400 images for the training data set, 70 for the validation data set, and 30 for the test data set.

6.5 Data Sets and Metrics

NSR accounts for a data set composed of 500 images taken from the *Imagenet* data set [48]. Given the ground-truth image **G**, NSR applies the artificial noise (e.g., Gaussian, speckle), generating the noisy image **N**. Then, the **N** image is down-sampled to the noisy low-resolution **L** image with a down-sampling factor. Two different down-sampling factors are applied: $k = 2$, where one row every two and one column every two is removed, and $k = 4$, where three columns and three rows every four are removed. Separated data sets are generated in terms of up-sampling factor and noise type, as the specialisation of the trained networks improves the accuracy of the reconstruction of the target image.

The quality preservation of the super-resolution is measured through quantitative metrics. The predicted image **P** is compared with **N** to measure the super-resolution accuracy and with **G** to measure the generated noise properties. The comparison accounts for the visual similarity of the super-resolution images (predicted and noisy) to quantify the preservation of the geometries and features and evaluate several metrics for local, structural, and texture similarity. Given the noisy and the predicted image on m points, we compute the quantitative metrics introduced in Sect. 2.4. Finally, we perform a qualitative assessment of blurring, artefacts, and noise patterns, and analyse the histogram properties of the noisy component with respect to the ground-truth image through $(\mathbf{P} - \mathbf{G})$ and compare with the histogram properties of the input noise $(\mathbf{N} - \mathbf{G})$.

6.6 Experimental Results

We test the super-resolution methods with 2X and 4X up-sampling factors on 2D images with Gaussian (Sect. 6.6.1), speckle noise (Sect. 6.6.2), and US 3D images (Sect. 6.6.3): the Gaussian noise has $\mu = 0$, $\sigma = 0.02$, and speckle noise has $\sigma = 0.02$. We compare NSR super-resolution with the methods CC [20], EDSR [27], SPM [43], SEI [50], MSSR [33], CLDSR [31], and CSNL [37].

6.6.1 Super-Resolution and Gaussian Noise

According to Tables 6.1 and 6.2, the NSR method has better results than previous work in terms of quantitative metrics. In particular, the average MSE value of NSR with a 2X up-sampling factor is 304, while all the other methods have an average MSE value of more than 400. SEI has a MSE value of 360. Also, the PSNR value of NSR with the 4X up-sampling factor is 20.6, while the best result of state-of-the-art methods is 20.1, performed by cubic convolution. In terms of structural metrics (i.e., SSIM and FSIM), NSR has better results than the other state-of-the-art methods in the 2X up-sampling. For example, the average SSIM of NSR is 0.831, while the best result of state-of-the-art methods is 0.823. In the 4X case, NSR has better results than other learning-based methods (i.e., EDSR and SPM), while other methods (e.g., CC and CSNL) have slightly better results. Finally, Fig. 6.2 shows the PSNR value box plot for the 2X and 4X up-sampling factor results. NSR improves the

Table 6.1 Concerning the 2X up-sampling factor results on Gaussian noise images, we report the metrics computed between target and super-resolution methods as average values on the test data set. The best results are in bold

Methods	CC	EDSR	SPM	CSNL	MSSR	CLDSR	SEI	NSR
MSE	406.76	809.74	683.21	421.62	521.05	1077	360.70	**304.11**
NRMSE	0.147	0.229	0.195	0.235	0.257	0.369	0.127	**0.116**
NCC	0.876	0.825	0.848	0.811	0.793	0.705	0.786	**0.891**
PSNR	23.09	20.47	21.78	16.88	15.96	12.36	21.01	**23.81**
SSIM	0.823	0.718	0.773	0.626	0.640	0.579	0.798	**0.831**
FSIM	0.914	0.860	0.886	0.546	0.559	0.500	0.781	**0.915**
UIQ	**0.998**	0.984	0.979	0.935	0.932	0.978	0.993	0.997

Table 6.2 Concerning the 4X up-sampling factor results on Gaussian noise images, we report the metrics computed between target and super-resolution methods as the average value on the test data set. The best results are in bold

Methods	CC	EDSR	SPM	CSNL	MSSR	CLDSR	SEI	NSR
MSE	1112	1880	1372	746	**674**	2940	677	949
NRMSE	0.299	0.381	0.331	0.241	0.237	0.464	**0.234**	0.277
NCC	0.822	0.721	0.780	0.880	**0.900**	0.652	0.891	0.843
PSNR	20.11	17.31	18.74	19.40	19.85	13.47	19.81	**20.64**
SSIM	0.591	0.473	0.553	**0.621**	0.616	0.595	0.605	0.564
FSIM	0.747	0.700	0.738	**0.777**	0.767	0.625	0.763	0.739
UIQ	0.992	0.989	0.988	0.992	0.992	0.987	0.993	**0.994**

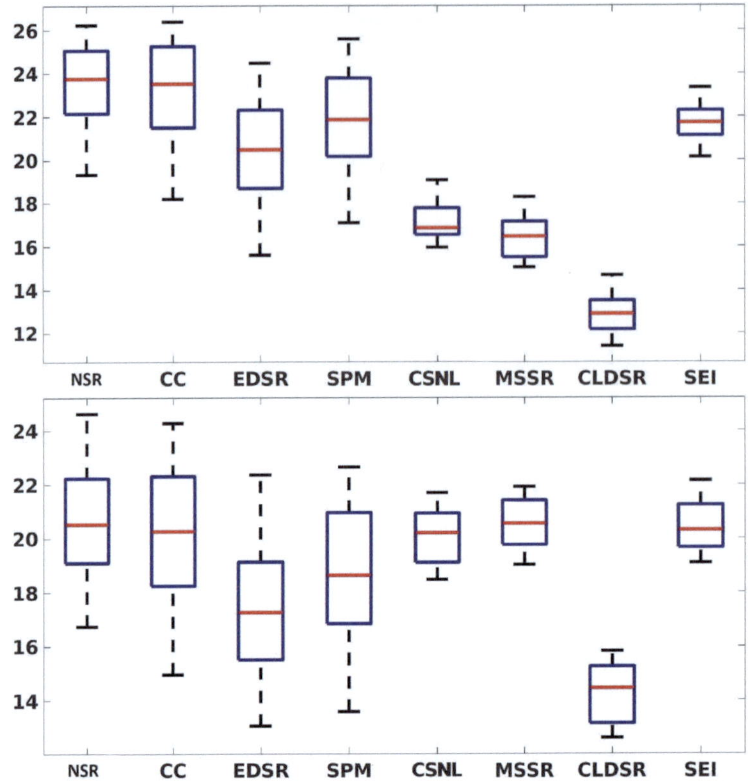

Fig. 6.2 2X (first row) and 4X (second row) box plot of PSNR with Gaussian noise. Image courtesy of [6]

average value, reduces the variability of the results of CC and SEI, and outperforms the results of the other methods, i.e., EDSR, SPM, CSNL, and CLDSR.

According to the histogram (Figs. 6.3, 6.4) of the generated noise of NSR versus state-of-the-art super-resolution methods, NSR better preserves the distribution of the Gaussian noise in terms of mean and standard deviation, both for the 2X and 4X up-sampling factors. CSNL and MSSR generate a noise distribution with a different mean value, while CLDSR does not preserve the standard deviation of the noise distribution of the input image.

6.6.2 Super-Resolution and Speckle Noise

Figure 6.5 shows the super-resolution results with speckle noise at 2X applied on a phantom image reproducing an ultrasound scanning on cysts at various dimensions. All the methods preserve the contours of the cysts. CLDSR generates a scattering artefact on the background

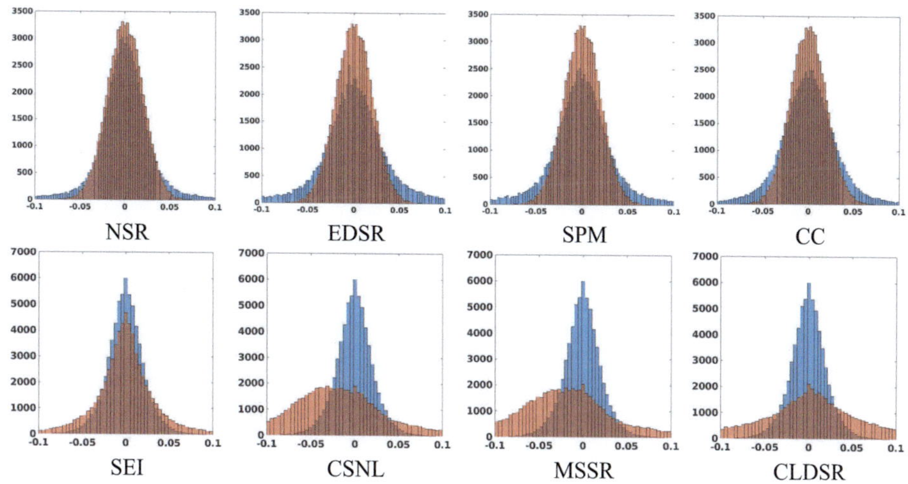

Fig. 6.3 2X up-sampling factor, Gaussian noise distribution: ground-truth (blue histogram) and super-resolution method (red histogram). Image courtesy of [6]

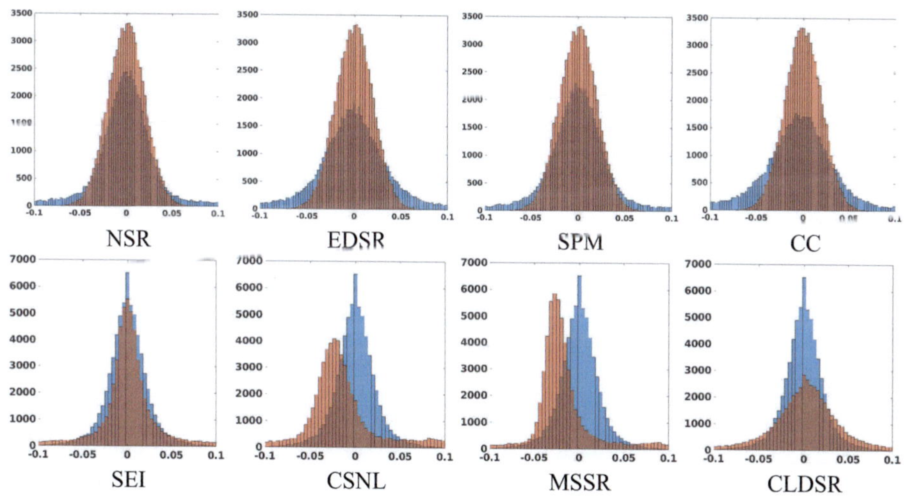

Fig. 6.4 4X up-sampling factor, Gaussian noise distribution: ground-truth (blue histogram) and super-resolution method (red histogram). Image courtesy of [6]

that affects the visual similarity of the prediction with the input image. Table 6.3 shows that NSR results are better than state-of-the-art super-resolution methods in quantitative metrics. In particular, the MSE value of NSR with a 2X up-sampling factor is 247, while all the other methods have a MSE value of more than 400. NSR has comparable results with respect to other learning-based techniques (i.e., EDSR and SPM) concerning SSIM and FSIM metrics, while CC has slightly better results. The MSSR method has worse results with respect to

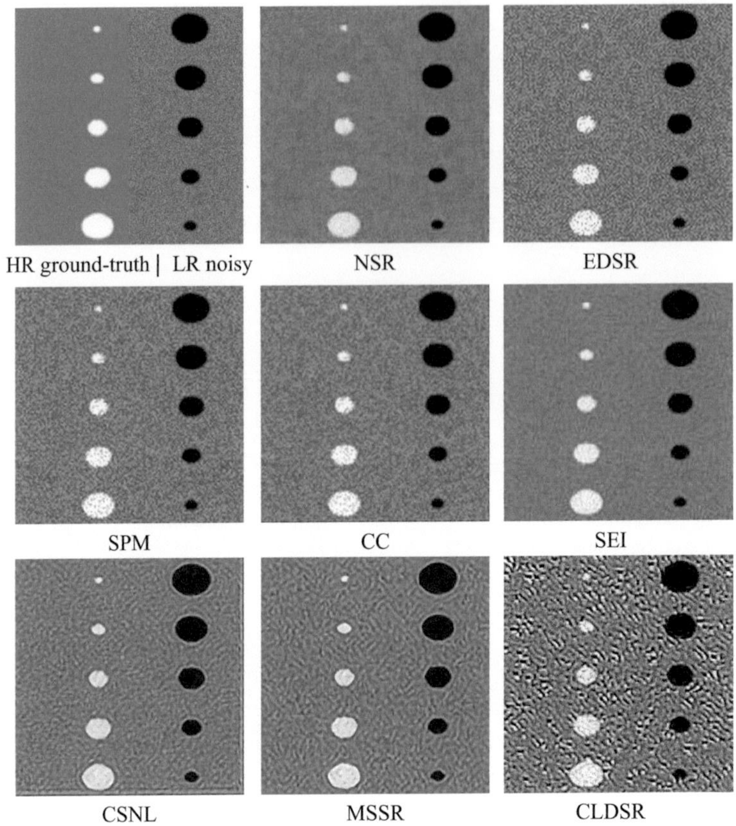

Fig. 6.5 Comparison among super-resolution methods, 2X up-sampling factor, speckle noise. Image courtesy of [6]

NSR in terms of PSNR (24.19 vs. 21.31) and FSIM (0.818 vs. 0.792). SEI has the worst results in quantitative metrics despite qualitatively and visually having similar results to NSR. CC has the worst result in preserving the noise distribution (Fig. 6.6), while NSR has similar results to SPM. MSSR has good results in terms of noise distribution, while CLDSR shows a long tail distribution as the generated noise has a larger variance than the input noise.

Speckle Noise in Breast Ultrasound Images

We apply NSR trained with synthetic speckle noise images on an ultrasound image from the breast anatomical district with 2X up-sampling (Fig. 6.7). Previous learning-based works generate artefacts, e.g., blurring in SPM and scattering in EDSR. CC and NSR have com-

Table 6.3 Concerning the 2X up-sampling factor results on speckle noise images, we report the metrics computed between target and super-resolution methods as the average value on the test data set. The best results are in bold

Methods	CC	EDSR	SPM	CSNL	MSSR	CLDSR	SEI	NSR
MSE	408.81	600.21	488.39	522.22	480.74	2209	4433	**247.67**
NRMSE	0.1832	0.220	0.199	0.205	0.197	0.391	0.773	**0.121**
NCC	0.812	0.744	0.780	0.749	0.771	0.580	0.120	**0.853**
PSNR	22.01	20.35	21.24	20.95	21.31	14.68	11.66	**24.19**
SSIM	**0.365**	0.301	0.318	0.237	0.302	0.288	0.078	0.312
FSIM	**0.875**	0.819	0.854	0.784	0.792	0.553	0.591	0.818
UIQ	0.988	0.982	0.985	0.977	0.977	0.966	0.670	**0.990**

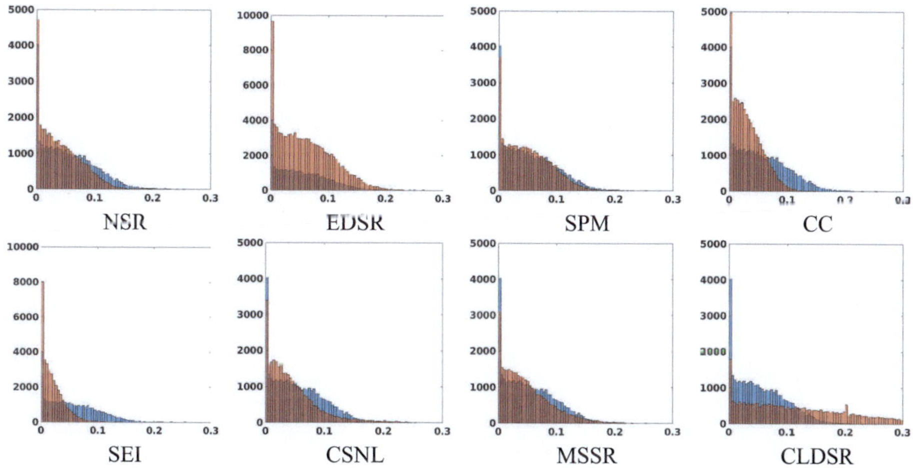

Fig. 6.6 2X speckle noise distribution: ground-truth (blue histogram) and super-resolution method (red histogram). Image courtesy of [6]

parable results, preserving textures and anatomical features without enhancing artefacts. SEI has comparable results with NSR and CC methods, while CSNL, MSSR, and CLDSR introduce artefacts and do not correctly preserve edges and anatomical features.

6.6.3 Super-Resolution of US 3D Images

We apply NSR trained with synthetic speckle noise images on ultrasound 3D images from the TUS-REC data set with 2X (Fig. 6.8) and 4X (Fig. 6.9) up-sampling for the reconstruction of non-acquired slices. NSR has better results than previous work with 2X up-sampling

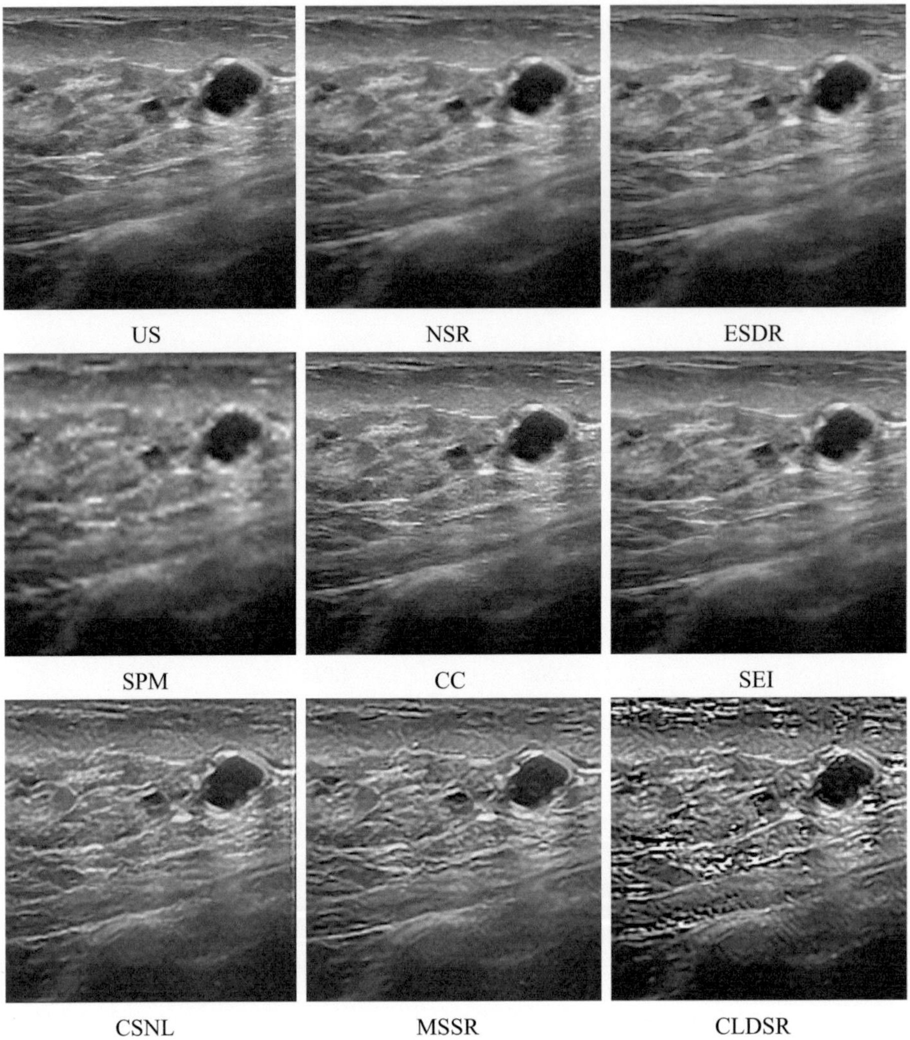

Fig. 6.7 Comparison among super-resolution methods on an ultrasound image from the breast district, 2X up-sampling factor. Image courtesy of [6]

in terms of preservation of textures, anatomical contours, and artefact reduction. EDSR generates a scattering effect on the skin, while SPM tends to smooth the prediction. CSNL, MSSR, and CLDSR do not correctly preserve anatomical contours. NSR has comparable results with CC with 4X up-sampling, with a slight scattering effect on the skin. The other methods generate artefacts; in particular, EDSR and SEI generate a strong scattering effect that alters the recognition of anatomical features.

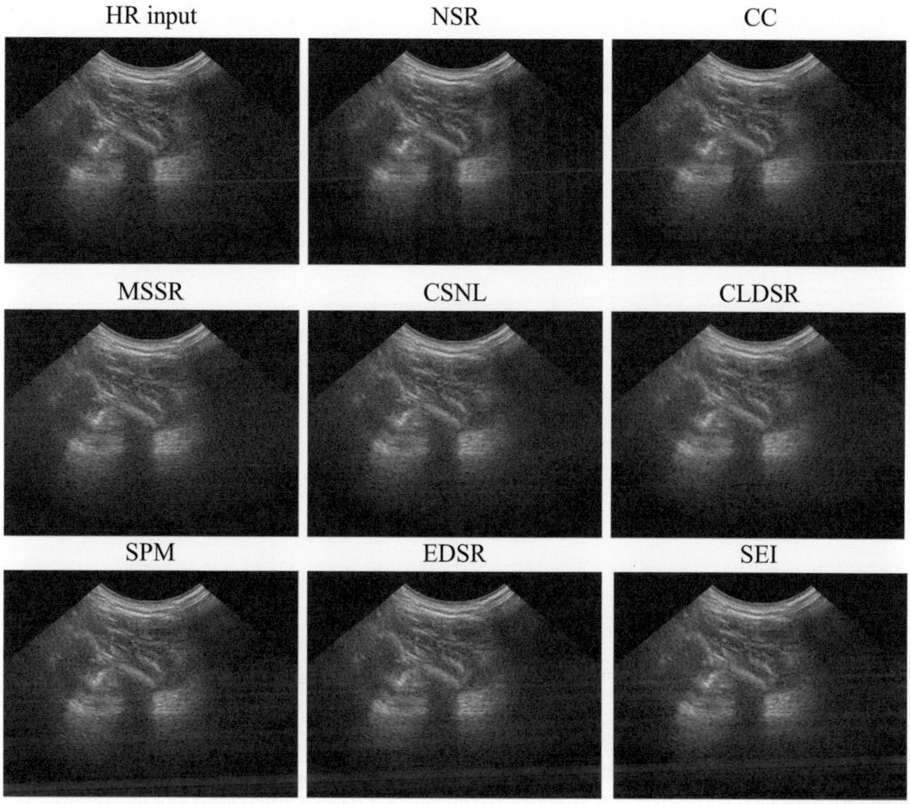

Fig. 6.8 Comparison of 2X super-resolution STAR methods, US 3D images from musculoskeletal anatomical district, right arm

Table 6.4 shows that NSR outperforms previous work in quantitative metrics. The MSE value of NSR 2X super-resolution is 5.05, while all the other methods have an MSE value greater than 12. NSR has the best results concerning feature-based (i.e., SSIM and FSIM) and noise-based (i.e., PSNR) metrics, with a PSNR value of 41.09 vs. 38.45 and UIQ value of 0.896 vs. 0.883. CSNL has the worst results in quantitative metrics; CC has poor results in terms of UIQ and SSIM metrics. Concerning the 4X up-sampling, the NSR method has the best results in terms of NRMSE (0.132), NCC (0.987), and PSNR (35.19) while showing a degradation in SSIM and UIQ metrics.

Training and Execution Time

Figure 6.10 reports the training and validation loss of the Gaussian 2X network, showing us the convergence of the NSR model. Concerning Eq. (6.1), the y-left axis shows the first term

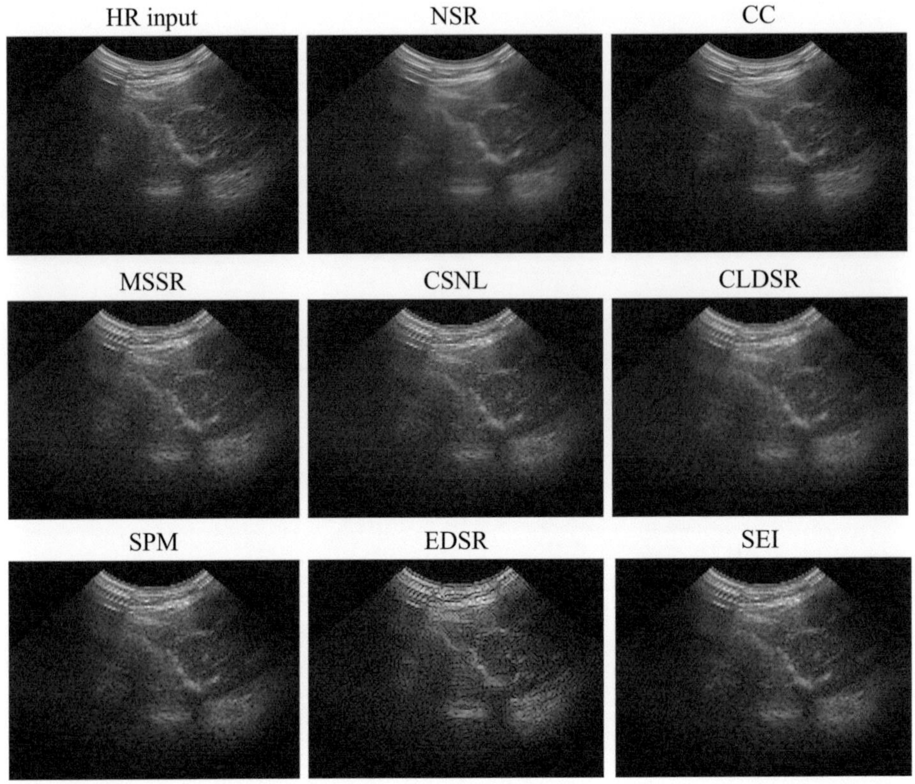

Fig. 6.9 Comparison of 4X super-resolution STAR methods, US 3D images from musculoskeletal anatomical district, left arm

of the loss (i.e., $\|\mathbf{P} - \mathbf{N}\|_F$), while the y-right axis shows the second term of the loss (i.e., the λ-weighted). NSR training minimises the first term (i.e., the approximation norm with respect to the noisy image) and contemporary increases the second term, i.e., the probability that the generated noise complies with the input noise distribution. We mention that the execution time of the training is around 30 s per epoch on the Cineca Marconi100 cluster, at the 26th position in the "*top500*" list [8]. The cluster uses 980 nodes, each with IBM Power9 AC922 at 3.1 GHz, 32 cores and 4 NVIDIA Volta V100 GPUs per node, with the GPU interconnection NVlink 2.0 at 16GB and 256GB of RAM each node. The prediction time is lower than 1 s on a standard workstation with 2 Intel i9-9900KF CPUs (3.60 GHz), 32GB RAM, and Tensorflow 2.7.

Execution Time of SR 3D US

We report in Fig. 6.11 the training and validation loss of the 2X SR model with respect to the number of epochs. Concerning Eq. (6.1), the y-left and right axes show the first

Table 6.4 Average metrics values computed between target and super-resolution methods applied to the TUS-REC data set. The best results are in bold

	Methods	CC	EDSR	SPM	CSNL	MSSR	CLDSR	SEI	NSR
2X	MSE	12.03	15.51	12.79	16.66	9.94	9.29	9.94	**5.05**
	NRMSE	0.103	0.118	0.107	0.122	0.094	0.091	0.095	**0.067**
	NCC	0.992	0.989	0.991	0.988	0.993	0.993	0.993	**0.997**
	PSNR	37.40	36.22	37.06	35.91	38.15	38.45	38.05	**41.09**
	SSIM	0.927	0.960	0.969	0.963	0.974	**0.975**	0.974	0.973
	FSIM	0.978	0.985	0.987	0.982	0.989	0.988	0.989	**0.989**
	UIQ	0.756	0.834	0.882	0.869	0.881	0.887	0.883	**0.896**
4X	MSE	22.46	76.57	47.48	62.46	46.96	41.33	45.79	**19.70**
	NRMSE	0.141	0.262	0.207	0.237	0.205	0.193	0.203	**0.132**
	NCC	0.984	0.948	0.967	0.957	0.967	0.971	0.968	**0.987**
	PSNR	34.61	29.29	31.36	30.18	31.41	31.97	31.52	**35.19**
	SSIM	0.877	0.840	0.897	0.878	0.896	**0.905**	0.898	0.786
	FSIM	0.974	0.910	0.937	0.930	0.936	0.943	0.938	**0.970**
	UIQ	**0.613**	0.456	0.602	0.572	0.600	0.612	0.604	0.438

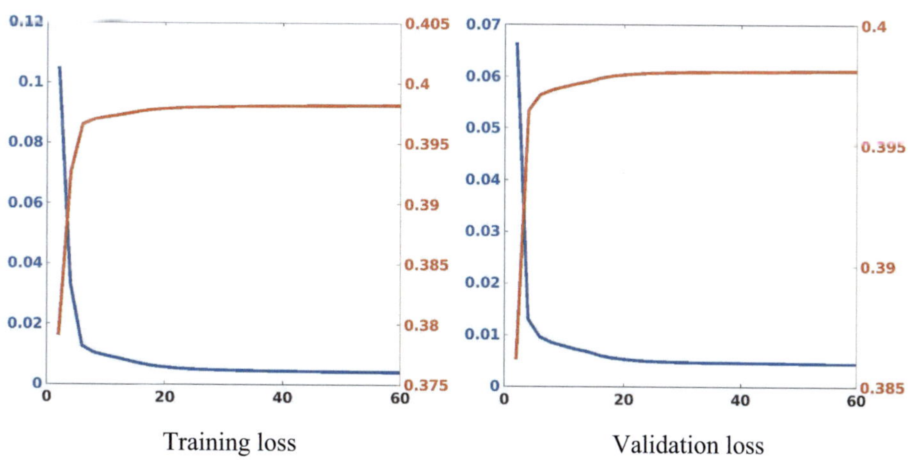

Fig. 6.10 Training and validation loss of the Gaussian 2X network: MSE loss (y-axis) with respect to the number of epochs (x-axis). Concerning Eq. (6.1), the y-left axis shows the first term of the loss, while the y-right axis shows the second term of the loss. Image courtesy of [6]

(i.e., $\|\mathbf{P} - \mathbf{N}\|_F$), and second (i.e., the λ-weighted) component of the loss, respectively. The training reduces the difference with respect to the target image (i.e., the first component of the loss). It increases the compliance of the generated noise with the speckle noise, i.e., the second component of the loss. The computation time of the training is 15 s for each epoch

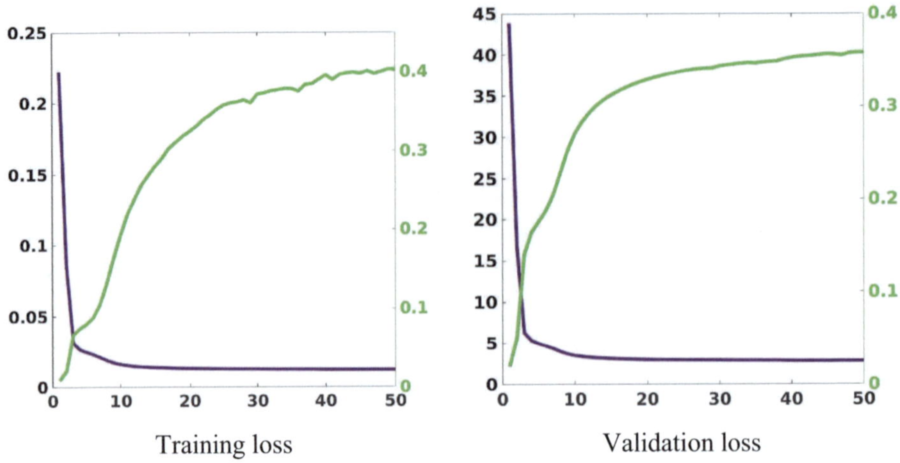

Fig. 6.11 Training and validation loss of the speckle 2X network: MSE loss (y-axis) with respect to the number of epochs (x-axis). Concerning Eq. (6.1), the y-left axis shows the first term of the loss, while the y-right axis shows the second term

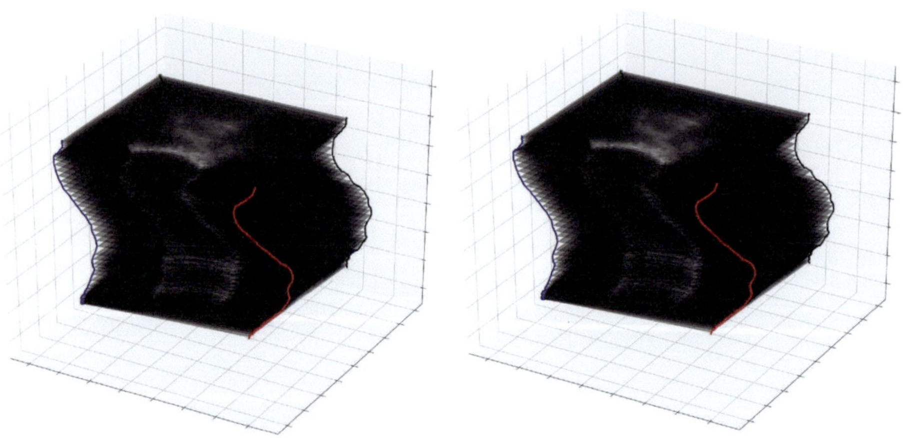

Fig. 6.12 Comparison between 3D reconstruction of HR input (left) and SR image (right)

on the CINECA Leonardo super-computer, currently at the 7th position in the "*top500*" list [8]. The prediction time for each slice is lower than 0.05 s on a workstation with 2 AMD Ryzen 9 7845HX CPUs (3.00 GHz) and 16GB RAM. As a preliminary result for the freehand 3D reconstruction, we report the 3D reconstruction [44] with HR input and SR images (Fig. 6.12).

References

1. M. Abdel-Nasser, O.A. Omer, Ultrasound image enhancement using a deep learning architecture, in *International Conference on Advanced Intelligent Systems and Informatics* (Springer, 2016), pp. 639–649
2. M. Abdel-Nasser et al., Breast tumor classification in ultrasound images using texture analysis and super-resolution methods. Eng. Appl. Artif. Intell. **59**, 84–92 (2017)
3. M. Alessandrini et al., A restoration framework for ultrasonic tissue characterization. IEEE Trans. Ultrason. Ferroelectr. Freq. Control **58**(11), 2344–2360 (2011)
4. O. Bar-Shira et al., Learned super resolution ultrasound for improved breast lesion characterization, in *International Conference on Medical Image Computing and Computer-Assisted Intervention* (Springer, 2021), pp. 109–118
5. A. Basarab et al., A method for vector displacement estimation with ultrasound imaging and its application for thyroid nodular disease. Med. Image Anal. **12**(3), 259–274 (2008)
6. S. Cammarasana, G. Patanè, Learning-based and quality preserving super-resolution of noisy images, in *Multimedia Tools and Applications* (2024), pp. 1–17
7. W. Choi et al., Deep CNN-based ultrasound super-resolution for high-speed high-resolution B-mode imaging, in *International Ultrasonics Symposium* (IEEE, 2018), pp. 1–4
8. *CINECA MARCONI100*. https://www.top500.org/system/179845/. Accessed 20 April 2022
9. G.T. Clement, J. Huttunen, K. Hynynen, Superresolution ultrasound imaging using back-projected reconstruction. J. Acoust. Soc. Amer. **118**(6), 3953–3960 (2005)
10. M. CüneyitoÄÅ¸lu Özkul, Ü.E. MumcuoÄÅ¸lu, Ä.T. Sancak, Single-image Bayesian restoration and multi-image super-resolution restoration for B mode ultrasound using an accurate system model involving correlated nature of the speckle noise. Ultrasonic Imaging **41**(6), 368–386 (2019)
11. K. Diamantis et al., Super-resolution axial localization of ultrasound scatter using multi-focal imaging. IEEE Trans. Biomed. Eng. **65**(8), 1840–1851 (2017)
12. J. Ding et al., Ultrasound image super-resolution with two-stage zero-shot CycleGAN. J. Phys.: Conf. Ser. **2031**(1), 012015 (2021)
13. C. Dong et al., Image super-resolution using deep convolutional networks. IEEE Trans. Pattern Anal. Mach. Intell. **38**(2), 295–307 (2015)
14. C. Dong et al., Learning a deep convolutional network for image super-resolution, in *European Conference on Computer Vision* (Springer, 2014), pp. 184–199
15. M.A. Ellis, F. Viola, W.F. Walker, Super-resolution image reconstruction using diffuse source models. Ultrasound Med. & Biol. **36**(6), 967–977 (2010)
16. S. Geman, D. Geman, Stochastic relaxation, Gibbs distributions, and the Bayesian restoration of images. IEEE Trans. Pattern Anal. Mach. Intell. **6**, 721–741 (1984)
17. K.T. Gribbon, C.T. Johnston, D.G. Bailey, A real-time FPGA implementation of a barrel distortion correction algorithm with bilinear interpolation. Image Vision Comput. New Zealand 408–413 (2003)
18. S. Han, T.B. Lee, Y.S. Heo, Deep image prior for super resolution of noisy image. Electronics **10**(16), 2014 (2021). ISSN: 2079-9292
19. J.-B. Huang, A. Singh, N. Ahuja, Single image super-resolution from transformed self-exemplars, in *Proceedings of the IEEE Conference on Computer Vision and Pattern Recognition* (2015), pp. 5197–5206
20. R. Keys, Cubic convolution interpolation for digital image processing. IEEE Trans. Acoust. Speech Signal Proce. **29**(6), 1153–1160 (1981)
21. D. Khaledyan et al., Low-cost implementation of bilinear and bicubic image interpolation for real-time image super-resolution, in *Global Humanitarian Technology Conference* (IEEE, 2020), pp. 1–5

22. P. Khavari, A. Asif, H. Rivaz, Non-local super resolution in ultrasound imaging, in *20th International Workshop on Multimedia Signal Processing* (IEEE, 2018), pp. 1–6
23. R. Kwok, W.T.K. Johnson, Block adaptive quantization of Magellan SAR data. Trans. Geosci. Remote Sensing **27**(4), 375–383 (1989)
24. A. Lapini, Comparison of super-resolution methods for quality enhancement of digital biomedical images, in et al., *8th International Symposium on Medical Information and Communication Technology (ISMICT)* (IEEE, 2014), pp. 1–5
25. R. Lavarello, F. Kamalabadi, W.D. O'Brien (2006) A regularized inverse approach to ultrasonic pulse-echo imaging. IEEE Trans. Med. Imag. **25**(6), 712–722 (2006)
26. C. Ledig et al., Photo-realistic single image super-resolution using a generative adversarial network, in *Proceedings of the IEEE Conference on Computer Vision and Pattern Recognition* (2017), pp. 4681–4690
27. B. Lim et al., Enhanced deep residual networks for single image super-resolution, in *Proceedings of the IEEE Conference on Computer Vision and Pattern Recognition Workshops* (2017), pp. 136–144
28. X. Lin et al., A super-resolution guided network for improving automated thyroid nodule segmentation. Comput. Methods Progr. Biomed. **227**, 107186 (2022)
29. F. Lingvall, A method of improving overall resolution in ultrasonic array imaging using spatio-temporal deconvolution. Ultrasonics **42**(1–9), 961–968 (2004)
30. D. Liu et al., Connecting image denoising and high-level vision tasks via deep learning. Trans. Image Proc. **29**, 3695–3706 (2020)
31. H. Liu et al., Fast simultaneous image super-resolution and motion deblurring with decoupled cooperative learning. J. Real-Time Image Proc. **17**, 1787–1800 (2020)
32. H. Liu et al., Perception consistency ultrasound image super-resolution via self-supervised CycleGAN. Neural Comput. Appl. 1–11 (2021)
33. H. Liu et al., Single image super-resolution using multi-scale deep encoder–decoder with phase congruency edge map guidance. Inf. Sci. **473**, 44–58 (2019). ISSN: 0020-0255. https://doi.org/10.1016/j.ins.2018.09.018. https://www.sciencedirect.com/science/article/pii/S0020025518307187
34. J. Lu, W. Liu, Unsupervised super-resolution framework for medical ultrasound images using dilated convolutional neural networks, in *3rd International Conference on Image, Vision and Computing* (IEEE, 2018), pp. 739–744
35. S. Maeda, Unpaired image super-resolution using pseudo-supervision, in *Proceedings of the IEEE/CVF Conference on Computer Vision and Pattern Recognition* (2020), pp. 291–300
36. G. Mahale et al., Hardware architecture of bi-cubic convolution interpolation for real-time image scaling, in *International Conference on Field-Programmable Technology* (IEEE, 2014), pp. 264–267
37. Y. Mei et al., Image super-resolution with cross-scale non-local attention and exhaustive self-exemplars mining, in *Proceedings of the IEEE/CVF Conference on Computer Vision and Pattern Recognition* (2020), pp. 5690–5699
38. R. Morin, A. Basarab, D. Kouamé, Alternating direction method of multipliers framework for super-resolution in ultrasound imaging, in *International Symposium on Biomedical Imaging* (IEEE, 2012), pp. 1595–1598
39. R. Morin et al., Post-processing multiple-frame super-resolution in ultrasound imaging, in *Medical Imaging: Ultrasonic Imaging, Tomography, and Therapy*, vol. 8320. (SPIE, 2012), pp. 433–440
40. Z. Mushtaq et al., Super resolution for noisy images using convolutional neural networks. Mathematics **10**(5), 777 (2022). ISSN: 2227-7390
41. M.K. Ng, P. Weiss, X. Yuan, Solving constrained total-variation image restoration and reconstruction problems via alternating direction methods. J. Sci. Comput. **32**(5), 2710–2736 (2010)

42. Y. Pang et al., JCS-Net: joint classification and super-resolution network for small-scale pedestrian detection in surveillance images. IEEE Trans. Inf. Forens. Secur. **14**(12), 3322–3331 (2019). https://doi.org/10.1109/TIFS.2019.2916592
43. T. Peleg, M. Elad, A statistical prediction model based on sparse representations for single image super-resolution. IEEE Trans. Image Proc. **23**(6), 2569–2582 (2014)
44. R. Prevost et al., 3D freehand ultrasound without external tracking using deep learning. Med. Image Anal. **48**, 187–202 (2018). ISSN: 1361-8415
45. K.G. Puschmann, F. Kneer, On super-resolution in astronomical imaging. Astron. & Astrophys. **436**(1), 373–378 (2005)
46. J. Qin et al., Progressive interaction-learning network for lightweight single-image super-resolution in industrial applications. IEEE Trans. Ind. Inf. **19**(2), 2183–2191 (2022)
47. N.C. Rakotonirina, A. Rasoanaivo, ESRGAN+: further improving enhanced super-resolution generative adversarial network, in *International Conference on Acoustics, Speech and Signal Processing* (IEEE, 2020), pp. 3637–3641
48. O. Russakovsky et al., Imagenet large scale visual recognition challenge. Int. J. Comput. Vision **115**, 211–252 (2015)
49. J. Salvador, E. Perez-Pellitero, Naive Bayes super-resolution forest, in *Proceedings of the IEEE International Conference on Computer Vision* (2015), pp. 325–333
50. J. Scanvic et al., *Self-Supervised Learning for Image Super-Resolution and Deblurring* (2023). arXiv:2312.11232 [eess.IV]
51. A. Singh, J. Singh, Survey on single image based super-resolution-implementation challenges and solutions. Multimedia Tools Appl. **79**, 1641–1672 (2020)
52. T. Taxt, R. Jirik, Superresolution of ultrasound images using the first and second harmonic signal. IEEE Trans. Ultrason. Ferroelectr. Freq. Control **51**(2), 163–175 (2004)
53. H. Temiz, H.S. Bilge, Super resolution of B-mode ultrasound images with deep learning. Access **8**, 78808–78820 (2020)
54. R. Timofte, V. De Smet, L. Van Gool, A+: adjusted anchored neighborhood regression for fast super-resolution, in *Asian Conference on Computer Vision* (Springer, 2014), pp. 111–126
55. R. Timofte, V. De Smet, L. Van Gool, Anchored neighborhood regression for fast example-based super-resolution, in *Proceedings of the IEEE International Conference on Computer Vision* (2013), pp. 1920–1927
56. S. Uchida, Image processing and recognition for biological images. Develop. Growth & Diff. **55**(4), 523–549 (2013)
57. R.J.G. Van Sloun et al., Deep learning for super-resolution vascular ultrasound imaging, in *International Conference on Acoustics, Speech and Signal Processing* (IEEE, 2019), pp. 1055–1059
58. A. Villar-Corrales, F. Schirrmacher, C. Riess, Deep Learning Architectural Designs for Super-Resolution of Noisy Images, pp. 1635–1639
59. F. Viola, M.A. Ellis, W.F. Walker, Time-domain optimized near-field estimator for ultrasound imaging: initial development and results. IEEE Trans. Med. Imaging **27**(1), 99–110 (2007)
60. C. Wang et al., A labelled ocean SAR imagery dataset of ten geophysical phenomena from Sentinel-1 wave mode. Geosci. Data J. **6**(2), 105–115 (2019)
61. X. Wang et al., ESRGAN: enhanced super-resolution generative adversarial networks, in *Computer Vision – ECCV 2018 Workshops*, ed. by L. Leal-Taixé, S. Roth (Springer International Publishing, Cham, 2019), pp. 63–79. ISBN: 978-3-030-11021-5
62. X. Wang et al., ESRGAN: enhanced super-resolution generative adversarial networks, in *Proceedings of the European Conference on Computer Vision* (2018)
63. Z. Wang, J. Chen, S.C.H. Hoi, Deep learning for image super-resolution: a survey. IEEE Trans. Pattern Anal. Mach. Intell. **43**(10), 3365–3387 (2020)

64. A. Webb, *Introduction to Biomedical Imaging* (2022)
65. G. Wu et al., *Transforming Image Super-Resolution: A ConvFormer-based Efficient Approach* (2024). arXiv:2401.05633 `[cs.CV]`
66. H. Xinong, Image denoising application in visual creation of art design based on intelligent algorithm. Soft Comput. 1–11 (2023)
67. J. Yang et al., Image super-resolution via sparse representation. IEEE Trans. Image Proc. **19**(11), 2861–2873 (2010)
68. Y.H. Yoon, J.C. Ye, Deep learning for accelerated ultrasound imaging, in *International Conference on Acoustics, Speech and Signal Processing* (IEEE, 2018), pp. 6673–6676
69. C. Yu, C. Zhang, L. Xie, An envelope signal based deconvolution algorithm for ultrasound imaging. In: *Signal processing* 92.3 (2012), pp. 793–800
70. J. Yu, Y. Fan, T. Huang, Wide activation for efficient image and video super-resolution, in *British Machine Vision Conference* (2020)
71. Y. Zhang et al., Single-image super-resolution based on rational fractal interpolation. IEEE Trans. Image Proc. **27**(8), 3782–3797 (2018)
72. N. Zhao et al., Joint Bayesian deconvolution and pointspread function estimation for ultrasound imaging, in *International Symposium on Biomedical Imaging* (IEEE, 2015), pp. 235–238
73. N. Zhao et al., Single image super-resolution of medical ultrasound images using a fast algorithm, in *International Symposium on Biomedical Imaging* (IEEE, 2016), pp. 473–476

7 Real-Time US Visualisation Based on Super-Resolution

We discuss a general deep-learning framework for the super-resolution of 2D US images (RT-SR) [1] by increasing the image resolution and reconstructing non-acquired scan lines. Applying the framework to US videos with a low spatial resolution and a high frequency (e.g., for the cardiac district), RT-SR generates high-frequency 2D US video with an increased spatial resolution of each frame, thus overcoming the main limits of current US probes, whose spatial resolution decreases as the acquisition frequency increases. To this end, RT-SR trains a neural network to improve the results of the up-sampling to match the target image (i.e., the high-resolution image). The resulting network does not perform the interpolation of the missing lines; this task is already performed by up-sampling. In contrast, the network learns how to transform the up-sampled lines into the target lines. To improve the quality of the up-sampling, RT-SR trains multiple networks; each one specialised to the input anatomical district (e.g., cardiac, abdominal) and its low-resolution image (e.g., 0.5X, 0.25X). This specialisation improves the quality of the up-sampling since we specialise the network to a specific prediction. The execution time of the super-resolution depends on the up-sampling and the network prediction; the prediction is achieved in real-time on standard medical hardware.

The *Wide Activation for Efficient and Accurate Image Super-Resolution* (WDSR) [9] is customised as a learning architecture (Sect. 7.1). The kernel size is selected according to the dimension of the low-resolution image to guarantee that at least two original lines (i.e., two lines acquired by the probe) are always included in the convolution operation. Then, the loss function can improve the prediction's visual accuracy. The logarithmic-based loss includes only up-sampled lines, excluding lines acquired by the probe. This approach is general in terms of the building blocks of the framework; in fact, we can select different up-sampling algorithms, e.g., *Single Image Super Resolution* (SISR) [5], *Enhanced Super Resolution Generative Adversarial Network* (ESRGAN) [8], and deep learning architectures,

S. Cammarasana and G. Patanè, *Real-Time Processing of Ultrasound Images*, Synthesis Lectures on Visual Computing: Computer Graphics, Animation, Computational Photography and Imaging, https://doi.org/10.1007/978-3-032-05805-8_7

e.g., *Pix2Pix* [2] and *VGG19* [7]. As experimental validation (Sect. 7.2), a quantitative and qualitative evaluation of the framework is performed on a large collection of US images acquired from different anatomical (e.g., musculoskeletal, obstetric, abdominal) districts.

7.1 Super-Resolution Framework

The framework for the real-time super-resolution of 2D images is composed of two steps: first, the up-sampling of the low-resolution image through an interpolating method; after the comparison of state-of-the-art methods (Sect. 6.6), the selection of *Cubic Convolution* as the up-sampling algorithm. Then, the application of a learning-based network to improve the visual accuracy of the up-sampling.

The US data set is considered for the tests, as it contains more than 10K US images at different resolutions and is acquired from different anatomical districts (e.g., obstetric, cardiac). Given a high-resolution image (i.e., the target) acquired by the probe, the corresponding low-resolution image is built by removing one line every 2 (0.5X) or 4 (0.25X). This approach is consistent with acquiring the US image, where the probe can acquire at the full, half, or a quarter of the maximum number of scan lines, depending on the activation of the piezoelectric crystals. The up-sampling of the low-resolution images through *Cubic Convolution* at 2X (applied to 0.5X low-resolution) or 4X (applied to 0.25X low-resolution) is performed. Then, the couple of up-sampled and target high-resolution images are used to analyse the proposed framework through the training and the prediction of the learning-based network, with a specialisation in anatomic districts.

RT-SR generates a separate data set of 1.5K images for each anatomical district and two different up-sampling resolutions of 2X and 4X. This approach requires the interpolation of the missing rows to the up-sampling method. At the same time, the learning model deals with predicting the target values from the interpolated values of the up-sampling method (Fig. 7.1). As *quantitative metrics*, we consider the PSNR for comparing the high-resolution target with the up-sampled image and the network prediction. We also compare the histogram of the absolute value of the prediction error to analyse the number of pixels whose error is lower than a certain threshold.

WDSR [9] architecture is selected as it exploits residual blocks and improves the prediction of images where the difference between the input and the target is small. RT-SR proposes an improved version of this network: *custom-WDSR*. Given an $y = L \times D$ target image, and its approximation $\hat{y}$, the loss function is defined as

$$Loss(y, \hat{y}) = \begin{cases} \sum_{l,d=1}^{L,D} \log \frac{|y_{ld} - \hat{y}_{ld}| + \epsilon}{k}, & \mathrm{mod}(l, s) = 0, \\ 0, \quad \text{otherwise}, \end{cases}$$

where s allows the network to exclude the lines acquired by the probe (e.g., $s = 4$ when 4X up-sampling is applied); $\epsilon = 10^{-4}$ avoids a null error for the logarithmic loss, and $k = 5$ determines the curvature of the logarithmic loss function. RT-SR enhances the pixels where

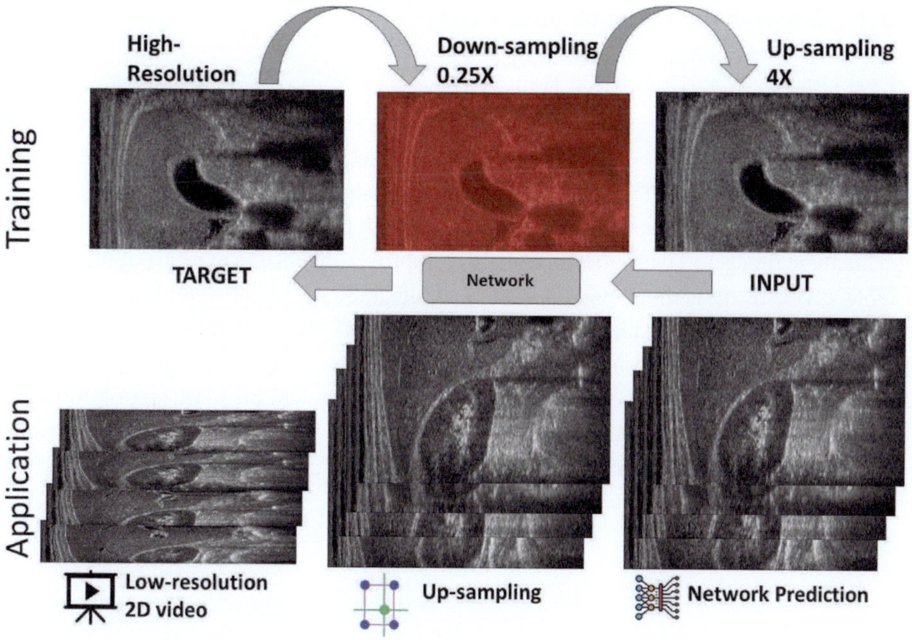

Fig. 7.1 Proposed framework: training of the learning-based model and spatial up-sampling of US videos. Image courtesy of [1]

the loss is less than 5 on the 0–255 range to improve the visual similarity between the prediction and the target image. The size of the kernel of the convolution filter depends on the up-sampling factor; in the case of a 2X up-sampling, we apply a 3×3 filter; for a 4X up-sampling, we apply a 5×5 filter. This choice allows the model to include at least two lines acquired by the probe in the convolution operator. Finally, the number of layers is set to 16 and the number of kernels to 10. The learning rate iteratively decreases, up to 10^{-6}, and the number of epochs is set to 200. The input and output layers of the network are $\#batch \times L \times D$ size.

7.2 Experimental Results

The real-time framework applies to US 2D images (Sect. 7.2.1) with an image-to-image learning model. Then, the application to each frame of a sequence of frames allows the user to increase the spatial resolution of a US 2D video (Sect. 7.2.2). Then, we discuss the execution time and computational cost (Sect. 7.2.3) of the framework. Finally, we review the super-resolution applied to previously denoised images (Sect. 7.2.4) and the comparison with previous work (Sect. 7.2.5).

7.2.1 Super-Resolution of US 2D Images

RT-SR trains each learning-based network (*custom-WDSR*) with 1.5K images, where the input is the outcome of the selected up-sampling method (i.e., *Cubic convolution*), and the target is the original high-resolution image. Input and target images have the same resolution, as the reconstruction of the missing lines has already been performed by *Cubic convolution*. Figures 7.2, 7.3 and 7.4 show the results of the network prediction, compared with the input and the target images. The framework visually improves the results in terms of blurring and artefacts. This result is more evident in the magnification of the ear of the foetus (Fig. 7.2), the mitral valve (Fig. 7.3), and the mass edges (Fig. 7.4). Figure 7.5(a-b-c, left) shows the box plot of the quantitative metrics, comparing the target images with the prediction and the *Cubic convolution*, respectively. The metrics are computed on a data set of 50 images of the same district and with the same up-sampling factor. We report that the PSNR median value improves by 1.5% on obstetric 4X images, 3.7% on cardiac 2X images, and 7.1% on abdominal 4X images.

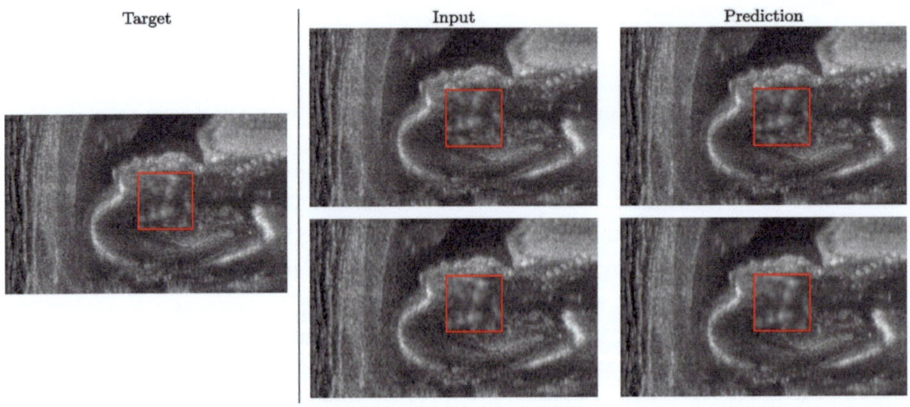

Fig. 7.2 Prediction on the raw images of the obstetric district: 2X up-sampling (first line); 4X up-sampling (second line). Image courtesy of [1]

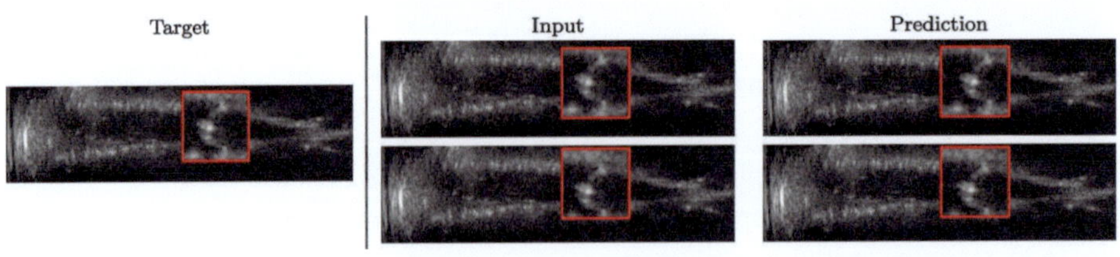

Fig. 7.3 Prediction on the raw images of the cardiac district: 2X up-sampling (first line); 4X up-sampling (second line). Image courtesy of [1]

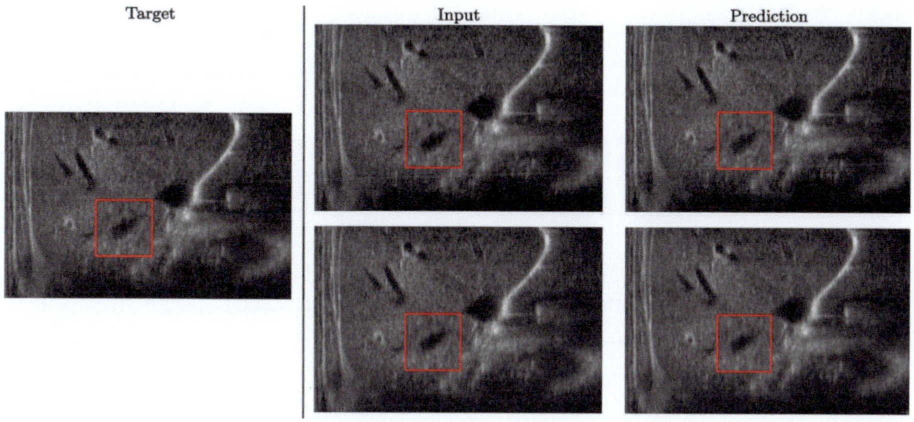

Fig. 7.4 Prediction on the raw images of the abdominal district: 2X up-sampling (first line); 4X up-sampling (second line). Image courtesy of [1]

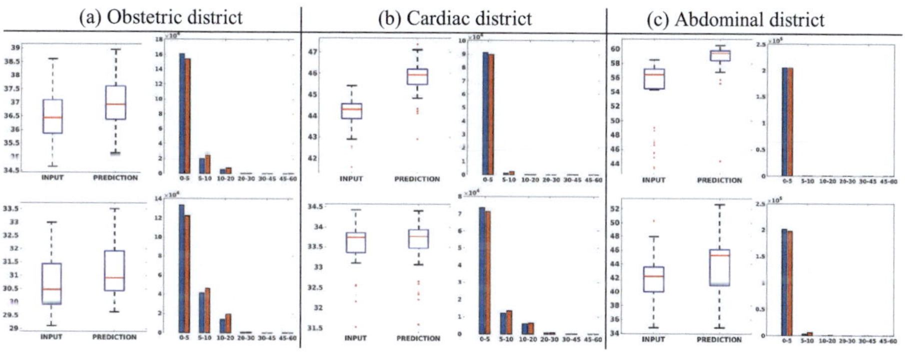

Fig. 7.5 PSNR box-plot district (left) of the **a** obstetric, **b** cardiac, and **c** abdominal districts, and error histogram (right): prediction (blue) versus input (red): 2X (first line) and 4X (second line) results. Image courtesy of [1]

Figure 7.5(a-b-c, right) shows the histogram of the absolute value of the error with respect to the target image, of the prediction and *Cubic convolution* results, respectively. The histograms show the number of pixels where the prediction error is lower than 5 (i.e., the first bin of the histogram). This means it is very similar to the target when visually analysing the images. From the *Cubic convolution* to the predicted images, this value increases of 9.0% on obstetric 4X images, 2.9% on cardiac 4X images, and 2.0% on abdominal 4X images. The analysis of the absolute value of the difference between the input and the prediction of the network (Fig. 7.6) shows that the alteration of the prediction to the pixel values ranges from 0 to a maximum absolute value of 20; furthermore, the black uniform areas are less affected by the prediction. Regarding the distance between the input and the prediction, we

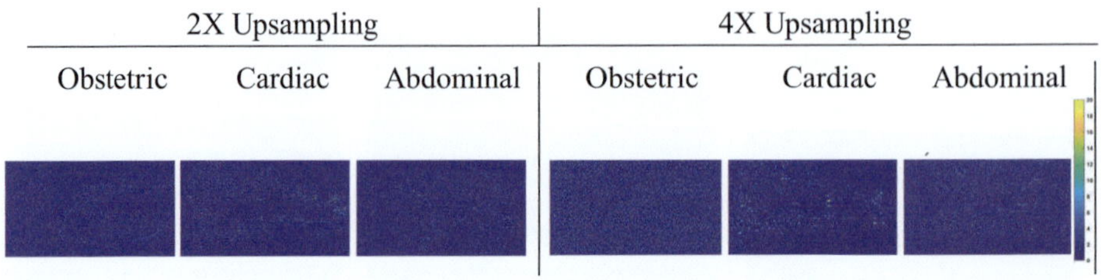

Fig. 7.6 Concerning Figs. 7.2, 7.3 and 7.4, we report the absolute value of the distance between the input and the prediction, for both 2X (first row) and 4X (second row) up-sampling factors. Image courtesy of [1]

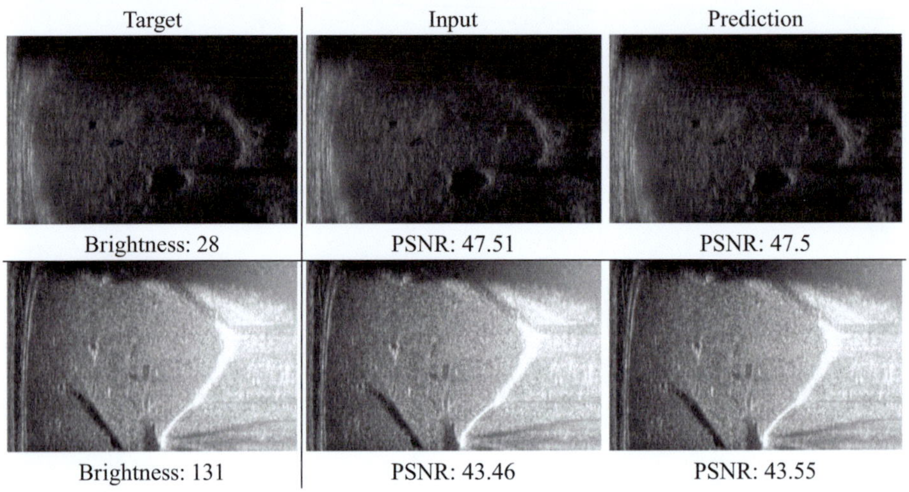

Fig. 7.7 Input and prediction of the raw images of the abdominal district 2X with different levels of brightness: low brightness (first row) and high brightness (second row). Image courtesy of [1]

do not observe a significant difference among anatomical districts and between 2X and 4X up-sampling.

We also verify the robustness of the RT-SR method on images at different brightness. Characterising the brightness of an image as the average value of all pixels, we test images with high and low brightness on different anatomical districts and up-sampling factors. Figures 7.7 and 7.8 show that the prediction performed with the trained network is robust to different image brightness values, never lowering the output accuracy or generating artefacts. Comparing the input and the prediction of the network with the target image, we improve the PSNR value from 43.46 to 43.55 with high brightness images from the abdominal district 2X up-sampling and from 31.01 to 31.48 with low brightness images from the obstetric district 4X up-sampling.

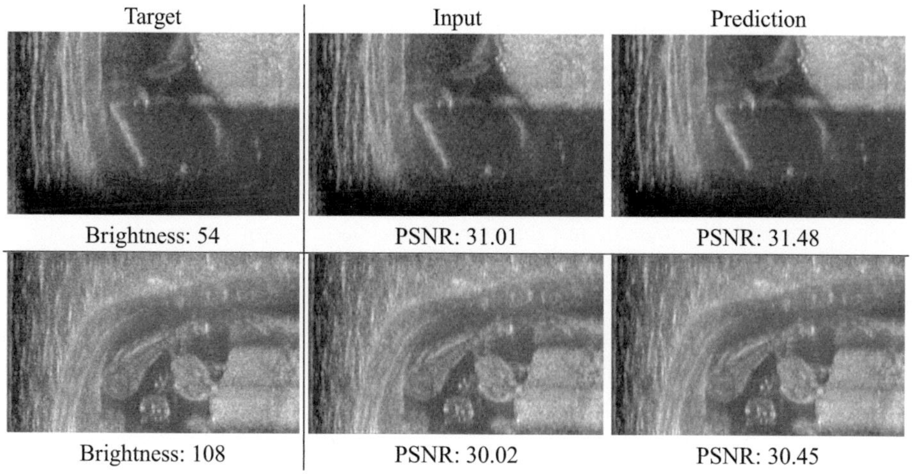

Fig. 7.8 Input and prediction of the raw images of the obstetric district 4X with different levels of brightness: low brightness (first row) and high brightness (second row). Image courtesy of [1]

7.2.2 Super-Resolution of US 2D Videos

Applying the approach to US videos with a low spatial resolution and a high frequency (e.g., for the cardiac district), high-frequency 2D US videos are generated with an increased spatial resolution of each frame, thus overcoming the main limits of current US probes, whose spatial resolution decreases as the acquisition frequency increases. The relationship between image resolution and video frequency f is given by $f = c/(2 \cdot d \cdot l)$, where c is the speed of sound. Acquiring low-resolution US images allows the physician to increase the acquisition frequency. The probe acquires fewer lines: we refer to 0.5X and 0.25X low-resolution images, as $l/2 \times d$ and $l/4 \times d$ resolution, respectively. In the video, the input signal is a 2D US video at full resolution $L \times D \times T$ with L lines, D depth, and T frames. Each image is down-sampled at $L/2$ or $L/4$, and we apply the framework for the spatial super-resolution to reconstruct the full-resolution 2D video.

7.2.3 Execution Time and Computational Cost

An HPC implementation of the framework on the CINECA-Marconi100 cluster is defined, exploiting both CPUs (IBM POWER9 AC922) and GPUs (NVIDIA Volta V100). Multiple networks are trained with large data sets for the target medical application, thus increasing the specialisation of anatomical districts and, consequently, the accuracy of the prediction. The training dataset can be periodically updated with the up-sampled images after the expert validation of the super-resolution results. To test the training phase of the learning-based networks in the HPC environment, 8 nodes are exploited, each composed of 32 cores and 4 accelerators, for a theoretical computational performance of 260 TFLOPS and 220 GB of memory per node.

The computational cost of the prediction depends on the resolution of the input image and the architecture of the network: in particular, the computational cost of a convolution operation is $O(r/s_r \cdot c/s_c) \cdot (f_r \cdot f_c) \cdot f$; in the application, the input images have variable resolutions, with a maximum value of $r = c = 600$, the kernel-filter size on rows and columns is $f_r = f_c = 3$ on 2X applications and $f_r = f_c = 5$ on 4X applications, the stride on rows and columns is $s_r = s_c = 1$, 16 convolution operators, and 10 kernel filters are applied. The prediction is tested on GPU-based hardware, replicating the hardware of a US scanner currently in use. Given a set of US input images from different districts at different resolutions, the average execution time is 14 ms.

7.2.4 Super-Resolution with Denoised Images

A soft denoising filter is applied to the input raw images to evaluate the effect of denoise pre-processing for the super-resolution of US images. In particular, the low-rank denoising method (Chap. 4) is applied, as it allows the user to select the intensity of the smoothing. This approach generates denoised images that are visually similar to raw images and are simultaneously more uniform. Then, down-sampled images are generated at 0.5X and 0.25X, and the *Cubic convolution* up-sampling is applied. These couple of images (i.e., denoised at full resolution and up-sampled) are used to train the learning-based network. With this approach, the performance of the up-sampling algorithm and the learning-based prediction, when applied to the input-denoised images, is verified.

Figure 7.9 shows the results of the prediction of the network, compared with the input and the target denoised images of the obstetric district. The framework visually improves the results in terms of blurring and artefacts. Figure 7.10 (left) shows the box plot of the quantitative metrics, comparing the target images with the prediction and the *Cubic convolution*, respectively. The PSNR metric is computed on a dataset of 50 images belonging to the same district and with the same up-sampling factor. Concerning the corresponding raw images (Fig. 7.5(a, left)), the denoising allows the network to improve the results of the up-sampling and the prediction significantly. In particular, comparing the target images with the predicted images, the median PSNR value of obstetric 2X denoised images is 51.8, compared to the median PSNR value of obstetric 2X raw images, which is 36.9.

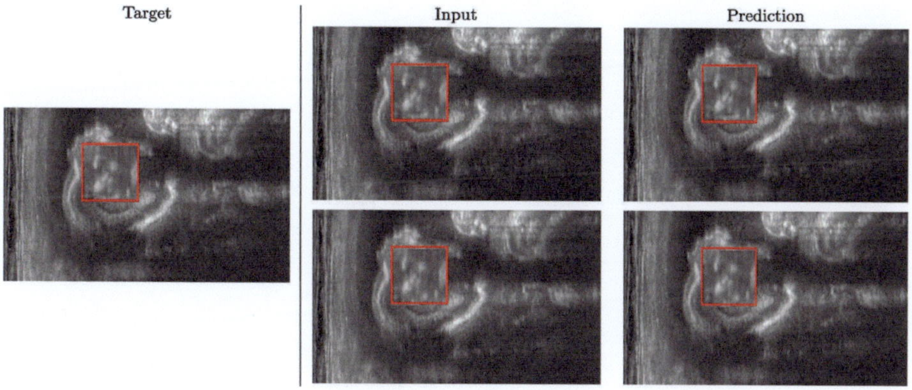

Fig. 7.9 Prediction on the denoised images of the obstetric district: 2X up-sampling (first line); 4X up-sampling (second line). Image courtesy of [1]

Figure 7.10 (right) shows the histogram of the absolute value of the error with respect to the target, of the prediction and *Cubic convolution* respectively. This result shows that the framework increases of 1.7% and 14% (2X and 4X, respectively) the number of pixels where the prediction error is lower than the defined threshold, and the prediction is very similar to the target when visually analysing the images, and improved compared to the learning framework applied to raw images. Finally, the denoise pre-processing can be performed in real-time through the learning-based denoising framework (Chap. 5).

7.2.5 Comparison with Previous Work

The comparison among state-of-the-art algorithms is used to select the up-sampling method of the framework, and the comparison of the results with previous work. Among up-sampling STAR methods, we test four methods belonging to different classes: *Cubic Convolution* [3], a kernel-based interpolating method; *Enhanced Deep Residual Networks—EDSR* [4], a learning-based method trained on generic images; *Enhanced Super-Resolution Generative Adversarial Network Plus—ESRGAN+* [6], a learning-based GAN method, specialised on US images with a dedicated training; *Single Image Super Resolution—SISR* [5], an up-sampling method which exploits sparse representations. We evaluate the up-sampling results of the selected methods on different anatomical districts and resolutions: obstetric district with 0.5X down-sampling (Fig. 7.11); abdominal district with 0.25X down-sampling (Fig. 7.12).

Table 7.1 summarises a comparison with the PSNR metric on a test data set of 50 images. *Cubic convolution* has a mean PSNR value of 36.52 and 42.17 for 2X and 4X upsampling, respectively. According to these results, *Cubic convolution* is selected as the best method for up-sampling US images. This method interpolates the missing lines without generating

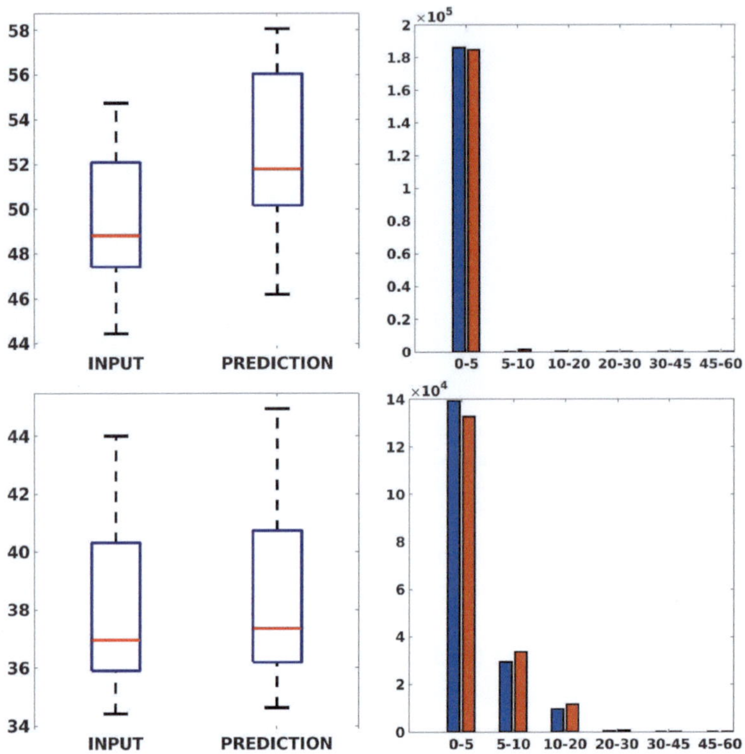

Fig. 7.10 PSNR box-plot of the obstetric district with denoised images (left) and error histogram (right): prediction (blue) versus input (red): 2X (first line) and 4X (second line) results. Image courtesy of [1]

artefacts. Also, *Cubic convolution* has a reduced computational cost ($O(kn)$, with n pixels) that well suits an application in a real-time context. In comparison, the RT-SR method improves the results of previous work (Figs. 7.11, 7.12, Table 7.1), with a mean PSNR value of 37.00 and 44.35 for 2X and 4X super-resolution, respectively. Finally, we underline that 4X super-resolution on the abdominal district has better results than 2X super-resolution on the obstetric district due to the complexity and variety of each anatomic district data set. Table 7.2 provides an overview of the compared super-resolution methods in terms of their main features (e.g., real-time, learning-based).

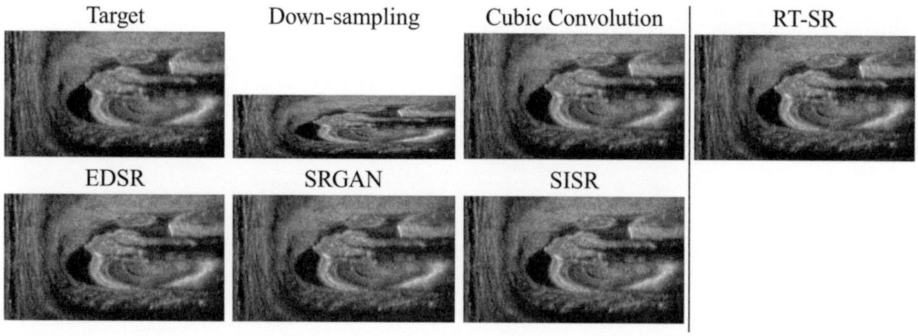

Fig. 7.11 Comparison of up-sampling methods versus RT-SR method on the obstetric district: 0.5X low-resolution and 2X up-sampling. See also Table 7.1. Image courtesy of [1]

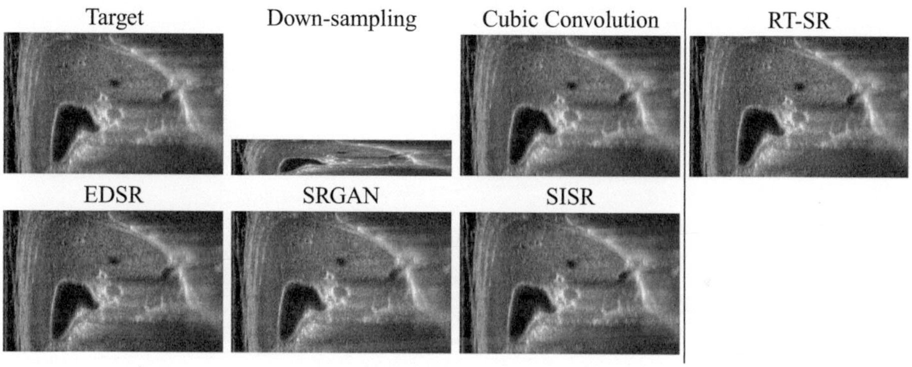

Fig. 7.12 Comparison of up-sampling methods versus RT-SR method on the abdominal district: 0.25X low-resolution and 4X up-sampling. See also Table 7.1. Image courtesy of [1]

Table 7.1 Concerning Figs. 7.11 and 7.12 we report the PSNR metric computed between the target and each up-sampling method as the mean value among the 50 test images

Test	Obstetric 2X	Abdominal 4X
Cubic Convolution	36.52	42.17
EDSR	32.08	34.91
SRGAN	33.70	36.35
SISR	34.75	38.58
RT-SR	**37.00**	**44.35**

Table 7.2 Comparison of super-resolution methods

Method	Real-time supe-res.	Learning-based approach	Up-sampling factor	Anatomical district	Noise management	3D US and/or temporal US
RT-SR	✓	✓	✓	✓	✓	✓
NSR	✓	✓	✗	✗	✓	✗
EDSR	✗	✓	✓	✗	✗	✗
SPM	✗	✓	✓	✗	✓	✗
CC	✗	✗	✓	✗	✗	✗
SEI	✗	✓	✓	✗	✓	✗
CSNL	✗	✓	✓	✗	✗	✗
MSSR	✗	✓	✓	✗	✓	✗
CLDSR	✓	✓	✓	✗	✗	✗

References

1. S. Cammarasana, P. Nicolardi, G. Patane, Super-resolution of 2D ultrasound images and videos. Med & Biolog. Eng. & Comput. **61**(10), 2511–2526 (2023)
2. P. Isola et al., Image-to-image translation with conditional adversarial networks. In: Proceedings of the IEEE Conference on Computer Vision and Pattern Recognition. 2017, pp. 1125 1134
3. R. Keys, Cubic convolution interpolation for digital image processing. IEEE Trans. Acous. Speech Signal Proc. **29**(6), 1153–1160 (1981)
4. B. Lim et al., Enhanced deep residual networks for single image superresolution, in *Proceedings of the IEEE Conference on Computer Vision and Pattern Recognition Workshops* (2017), pp. 136–144
5. T. Peleg, M. Elad, A statistical prediction model based on sparse representations for single image super-resolution. IEEE Trans. Image Proc. **23**(6), 2569–2582 (2014)
6. N.C. Rakotonirina, A. Rasoanaivo, ESRGAN+: further improving enhanced super-resolution generative adversarial network, in *International Conference on Acoustics, Speech and Signal Processing* (IEEE, 2020), pp. 3637–3641
7. K. Simonyan, A. Zisserman, Very deep convolutional networks for large-scale image recognition (2014). arXiv:1409.1556
8. X. Wang et al., ESRGAN: enhanced super-resolution generative adversarial networks, in *Proceedings of the European Conference on Computer Vision* (2018)
9. J. Yu, Y. Fan, T. Huang, Wide activation for efficient image and video super-resolution, in *British Machine Vision Conference* (2020)

Future Technological & Research Perspectives 8

We discuss future work in terms of analysis and development of novel methods for the further processing and analysis of US signals, such as AI-driven segmentation, morphological and predictive analysis, the application of microbubbles for US super-resolution, and research perspectives.

AI-Driven Processing and Analysis of US Images

Deep learning is widespread in analysing US images for lesion/nodule classification, organ segmentation, and mass detection. The application of deep learning and task-specific clinical knowledge [17] supports the automatic detection and classification of thyroid nodules with different types of nodule diagnoses and achieves better performances than experienced radiologists; in [14], the presence of metastatic lymph nodes is evaluated through DL techniques as a prognostic indicator for thyroid carcinomas. Several works apply learning models to classify tumours; transfer learning and colour conversion [7] classify tumours, comparing the results with the assessment of experienced radiologists. The segmentation of the *region of interest* (ROI) is also crucial to help physicians and processing algorithms to evaluate and classify tumours; Sobel operators are applied to obtain a gradient magnitude image [12] to classify breast tumours into three major types: cyst/mass, fatty tissue, and fibro-glandular tissue.

In the obstetric anatomical district, several clinical assessments are performed through AI techniques, such as respiratory morbidity [5], identification and classification of foetal standard planes [30], and prediction of follicles' maturity during the reproductive cycle [15]. Other applications of deep learning in US concern texture analysis for the prediction of hypertension in pregnancy [13], quantitative analysis for evaluating COVID-19 pneumo-

S. Cammarasana and G. Patanè, *Real-Time Processing of Ultrasound Images*, Synthesis Lectures on Visual Computing: Computer Graphics, Animation, Computational Photography and Imaging, https://doi.org/10.1007/978-3-032-05805-8_8

nia [8] and spontaneous preterm birth [3], co-registration with 3D MRI and CT [27], and real-time guidance for screw insertion for scaphoid bone fixation after a fracture [1]. Applying deep learning methodologies in ultrasound processing requires attention to the design of the training data set, metrics, and learning architecture to prevent the generation of patterns/artefacts. For a further review of deep learning methods for ultrasound processing, we refer the readers to [16, 24].

In this context, US imaging is applied for:

- *supplemental screening* for women with dense breast tissue, where traditional mammograms may not detect tumours. Ultrasound can dense tissues and improve the accuracy of screening;
- *cysts and solid masses identification* where US image allows the clinician to perform a less invasive procedure (e.g., biopsies). However, we mention that US imaging is less effective than mammography in detecting microcalcifications that may indicate early tumours; and
- *biopsy guidance* where ultrasound imaging allows the surgeon to guide needle biopsies.

Advanced image processing and segmentation are widespread to distinguish between malignant and benign tumours in several anatomical districts, e.g., breast, abdomen, and to identify the tumour from surrounding tissues. Tumour segmentation can be affected by tumour appearance, the presence of noise, low resolution of the image, and overlapping of image intensity between tumour and background.

Region-based methods (e.g., region growing, watershed segmentation) apply pixel connectivity to segment the tumour while requiring fine-tuning to handle noise and intensity variation. Canny edge detector and Sobel filters perform edge detection, finding the changes in the magnitude of pixel intensity, which indicates the presence of a boundary. Learning-based methods (e.g., variational autoencoders (VAEs) and generative adversarial networks (GANs)) are widespread in many anatomical districts with relevant results. However, these methods require a large data set for training and may generate artefacts in anatomical structures. We refer the reader to surveys on breast [28] and thyroid [29] tumour segmentation. Some preliminary results on breast image segmentation with a GAN method are presented in Fig. 8.1.

Once the tumour is segmented, the classification determines if the tumour is malignant or benign or tumour subtypes for personalised treatment planning. First, several features, such as shape, texture, and intensity, are extracted from the segmented tumour region. For the classification part, both traditional classifiers (e.g., support vector machine (SVM) [22] and random forest [26]) and learning-based methods [4] are applied Given the lack of annotated ultrasound datasets, transfer learning is commonly used in breast tumour classification [2] to improve the accuracy of the model. The performance of the segmentation and classification models is measured through accuracy, sensitivity, specificity and dice similarity coefficient (DSC) metrics. DSC measures the overlap between the predicted tumour region and the

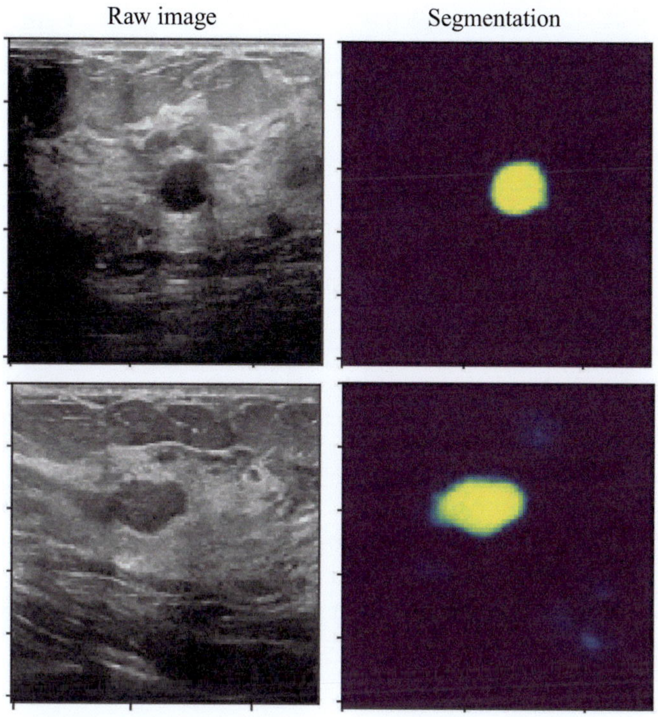

Fig. 8.1 Raw image (left) and segmentation (right) of benign (first row) and malignant (second row) tumours from breast ultrasound images

ground truth. High sensitivity and specificity are required for classification models to not misdiagnose malignant tumours. Ultrasound imaging is widespread for early detection of breast [20], pelvis [11], and kidney [23] tumour diagnosis. Some preliminary results on feature extraction from segmented images and classification into benign versus malignant breast tumours with the SVM method are presented in Fig. 8.2.

As the main challenges of US breast tumour segmentation and classification, speckle noise affects the visualisation of the tumour boundaries and reduces the accuracy of the segmentation. Low contrast and artefacts, especially in dense breast tissue, affect the differentiation of tumours from normal tissue. The limited field of view of the US reduces the portion of the image acquisition, increasing the chance of missed lesions in peripheral regions. Interoperator variability affects the quality and consistency of the labelled datasets used to train learning-based models. Finally, boundary ambiguity and blending into surrounding tissues affect the delineation of tumour boundaries.

In this context, 3D ultrasound images provide a volumetric view of the breast with better spatial information and more accurate tumour volume assessment. Automated breast ultrasound applies an automated transducer to scan the whole breast, overcoming the operator

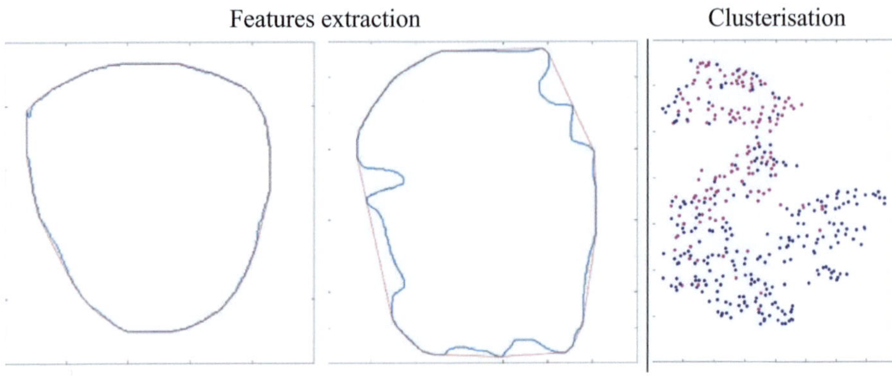

Fig. 8.2 Features extraction of benign (left) and malignant (centre) tumours; classification of benign (blue) and malignant (magenta) tumours

dependency of traditional ultrasound. Elastography is an ultrasound technique that measures tissue stiffness. As malignant tumours are stiffer than benign masses, elastography provides more information about the tumour characteristics and improves the accuracy of the classification.

Microbubbles for US Super-Resolution

Microbubbles are injected intravenously as a contrast agent and flow within the entire vasculature [6, 10], acting as resonators with a resonance frequency of 1–15 MHz and increasing their scattering coefficient in the clinical frequency range. Microbubbles re-emit ultrasound nonlinearly to highlight the vasculature and capillaries. In contrast to optical agents, microbubbles are entirely intravascular due to their micrometre size and can be detected deep within the body. However, ultrasound remains limited by resolution. *Ultrasound localisation microscopy* (ULM) is a super-resolution imaging technique that accounts for the localisation and tracking of intravascular *microbubble* (MB) contrast agents to reconstruct microvasculature. The ULM MB localisation achieves a trade-off between imaging resolution, which is imposed by wavelength, and imaging penetration depth, which is limited by attenuation. High-fidelity MB localisation requires non-overlapping MB signals, while the dilution of MBs used to achieve spatially sparse signals requires a long imaging duration to ensure the full perfusion of all patent vasculature. Due to long imaging duration, the image quality is affected by tissue motion and time-varying changes in vascular flow. ULM requires a correct concentration of the injected microbubbles, and increased depth reduces the performance of ultrasound super-resolution. A low concentration generally requires a longer acquisition time, and a high concentration increases the probability of more than one microbubble in the same imaging resolution cell, generating noisy images due to the presence of multiple detected signals.

Previous work on super-resolution with acoustically activated nanodroplets is focused on generating super-resolved micro-vessel maps and not producing blood flow velocity maps. In [18, 25], neural network architectures have been applied to extract point target locations representing MB centroids from B-mode images in the spatial domain. In [19, 21], spatio-temporal data and a fully convolutional network have been applied to improve localisation confidence. In [9], *Deep-SMV* is trained to generate super-resolved microvascular structural and velocity maps by identifying diffraction-limited MB signals, ranging from spatially sparse to overlapping conditions. *Sparsity-based and deep learning-based approaches* overcome this constraint to extract useful vascular structural information from overlapping microbubble signals and apply a structurally sparse prior-representing vasculature to resolve overlapping MB point-spread functions (PSF). *Compressive sensing algorithms* resolve high-density MB localisations, and deconvolution-based approaches iteratively shrink the imaging PSF, improving the localisation accuracy for overlapping MB signals. Learning-based super-resolution micro-vessel velocimetry (Deep-SMV) is a localisation-free approach that provides high imaging speed and robustness to high MB concentrations and directly outputs blood velocity measurements at a super-resolution without explicit data association for MB tracking.

Research Perspectives

Current trends in ultrasound imaging aim to improve several aspects of ultrasound signal processing and analysis, such as segmentation, morphological and quantitative analysis, predictive modelling, and graph-based applications. *Segmentation* of anatomical structures differentiates between various typologies of tissue and is relevant for accurate diagnosis and treatment planning. Both standard and deep learning techniques are applied to reduce the manual intervention for clinicians while increasing the accuracy of the segmentation. *Morphological and quantitative analysis* of ultrasound images provides insights into the shapes and structures of tissues. This analysis is relevant for the evaluation of tumours or organ deformities and supports the physicians in the diagnosis. *Predictive analysis* is applied to US images to improve clinical decision-making in terms of the progression of diseases, management strategies, and drug administration. The application of *graph theory* in ultrasound signal processing supports the analysis of complex relationships between different anatomical structures, as the representation of anatomical features as nodes in a graph to identify the connectivities between anatomical structures and characterise diseases and co-morbidities. In this context, ethical aspects are relevant, as the integration of artificial intelligence into clinical practice implies data privacy, patients' consent, and bias in predictions, with a focus on patient privacy, disparities in healthcare outcomes, and representation of diversity in the population.

Future research directions include several aspects in terms of hardware, software, industrial and clinical applications. *Personalised medicine* allows physicians to provide patients with more targeted diagnostics and treatments and reduce unnecessary interventions. AI and ultrasound imaging provide insights into disease progression, treatment response and optimal treatment strategies based on individual characteristics. For example, in oncology, AI-enhanced ultrasound can assess tumour characteristics and vascularisation patterns to support treatment plans. In cardiology, personalised echocardiograms with AI-based quantification of heart function support decisions on valve repair and replacement. Personalised ultrasound is applied in obstetrics, where predictive methods optimise the monitoring of foetal growth and development. The analysis of large datasets of historical ultrasound images through machine learning can identify patterns that predict future health outcomes. As a matter of example, predictive analytics can forecast the progression of a certain aided management disease, readmission to hospitals, and the risk of complications during patients' surgery.

Multimodal imaging and fusion combine ultrasound with other imaging modalities (CT, MR, PET) to provide a complete view of the anatomy and pathology of the patient and support more accurate diagnoses and better treatment planning. For example, US-MRI fusion can give detailed soft tissue contrast and real-time dynamic imaging, supporting procedures such as tumour ablation where MRI provides detailed visualisation of the target tissue and ultrasound guides real-time intervention. Similarly, the US-PET combination enhances the detection of metabolic activity in tissues and provides both anatomical and functional information, which is critical in tumour diagnosis and treatment planning.

Point of Care US (POCUS) is an advanced diagnostic ultrasonography performed in real-time and interpreted by the attending physician at a bedside test (e.g., for the evaluation of pulmonary conditions, the screening of abdominal aortic aneurysms in the emergency department, the diagnosis of deep venous thrombosis). The use and growth of POCUS as a diagnostic, procedural, and screening tool have been possible thanks to the more compact, higher-quality, and affordable price of US devices. Point-of-care US is fundamental to improving diagnostic accuracy and safety during medical procedures, thus transitioning US examinations from secondary to primary care. Results from the secondary healthcare sector show that POCUS applications can improve the diagnostic process and achieve earlier diagnosis, with clear benefits for the patients and the use of healthcare resources, and reduce the need for patient evaluation by a second clinician. Healthcare applications benefit from secure, highly performant point-of-care computing technologies, tools, and devices for the processing and analysis of large sets of real-time data at the point of care (e.g., US images). Furthermore, the combination of point-of-care technologies with extended reality and visualisation techniques enables continuous and real-time monitoring in clinical settings and workflows. Physicians address person-centred health care by physically placing computation and storage adjacent to where data is created to improve patient safety in healthy living support, remote patient monitoring, surgery workflows, and acute care.

The diffusion of *portable and wearable ultrasound* devices improves ultrasound accessibility, accuracy, and versatility to support remote patient monitoring and personalised medicine. The miniaturisation of ultrasound devices affects point-of-care diagnostics: portable ultrasound scanners allow physicians to acquire images at the bedside, in rural clinics or in patients' homes. Wearable ultrasound allows patients to continuously monitor, such as ultrasound patches that monitor heart function in patients with cardiovascular disease or track muscle healing in athletes recovering from injuries. The analysis of these measures through AI methodologies allows the users to monitor and detect complications in advance. Portable ultrasound with *telemedicine* can extend high-quality imaging to remote areas where healthcare is limited, such as in developing countries where trained radiologists or other specialists are not available, thus connecting local clinicians with experts from around the world and reducing the need for expensive transfers to distant hospitals.

Ultrasound acquisition is operator-dependent, and the quality of the image is affected by the expertise and skills of the user. AI affects the *training and education* of healthcare providers in ultrasound, supporting trainees with real-time feedback, guiding the position of the ultrasound probe, adjusting the parameters of the image acquisition, and helping with the interpretation of the acquired images. Virtual and augmented reality create realistic ultrasound training simulations, supporting the operators with the acquisition and interpretation of the ultrasound images in a controlled and risk-free environment.

In this context, it is relevant to (i) develop and test innovative tools, devices and systems for the real-time processing and analysis of US images; (ii) provide clinicians with real-time imaging, analysis and interactive visual presentation for understanding and diagnosing diseases, facilitating risk assessment, prevention, and providing medical interventions with improved patient safety; (iii) integrate technologies from proof-of-concept to prototype demonstration in an operational environment (e.g., emergency room, radiology department) and the involvement of clinicians and manufacturers of medical devices, thus ensuring that the developed prototypes are tested and seamlessly integrated into existing digital infrastructures and clinical workflows (e.g., surgery workflows). As the main challenges, we mention:

- *Data privacy and security*: The protection of patient data is addressed through data encryption, data sharing protocols, Health Insurance Portability and Accountability Act (HIPAA), and AI act compliance to protect patient data.
- *Clinical validation*: AI algorithms require clinical validation in terms of image outcomes, generation of artefacts, and preservation of anatomical features. AI models have to be tested with different patient populations, across different conditions, and in different clinical environments to ensure generalisability and accuracy.
- *Interoperability*: Integrating AI methods with existing healthcare infrastructure (EHRs, hospital information systems) requires seamless interoperability through the standardisation of data formats and the assurance of compatibility between different ultrasound devices and AI platforms.
- *Ethical considerations*: The inclusion of biodiversities and the differentiation of the population within large and heterogeneous data sets to avoid biases in AI methods.

- *AI training*: The training of providers both in ultrasound acquisition and AI methodologies to interpret image outcomes and incorporate AI recommendations into clinical workflow.

References

1. E.M.A. Anas et al., Bone enhancement in ultrasound using local spectrum variations for guiding percutaneous scaphoid fracture fixation procedures. Int. J. Comput. Ass. Radiol. Surg. **10**(6), 959–969 (2015)
2. G. Ayana, K. Dese, S.-W. Choe, Transfer learning in breast cancer diagnoses via ultrasound imaging. Cancers **13**(4), 738 (2021)
3. N. Baños et al., Quantitative analysis of cervical texture by ultrasound in mid-pregnancy and association with spontaneous preterm birth. Ultrasound Obstetr. & Gynecol. **51**(5), 637–643 (2018)
4. A.S. Becker et al., Classification of breast cancer in ultrasound imaging using a generic deep learning analysis software: a pilot study. The Brit. J. Radiol. **91**(1083), 20170576 (2018)
5. X.P. Burgos-Artizzu et al., Evaluation of an improved tool for non-invasive prediction of neonatal respiratory morbidity based on fully automated fetal lung ultrasound analysis. Sci. Rep. **9**(1), 1–7 (2019)
6. P.N. Burns, S.R. Wilson, Microbubble contrast for radiological imaging: 1. Principles. Ultrasound Q. **22**(1), 5–13 (2006)
7. M. Byra et al., Breast mass classification in sonography with transfer learning using a deep convolutional neural network and color conversion. Med. Phys. **46**(2), 746–755 (2019)
8. J. Chen et al., Quantitative analysis and automated lung ultrasound scoring for evaluating COVID-19 pneumonia with neural networks. IEEE Trans. Ultrason. Ferroelectr. Freq. Control **68**(7), 2507–2515 (2021)
9. X. Chen et al., Localization free super-resolution microbubble velocimetry using a long short-term memory neural network. Trans. Med. Imag. **42**(8), 2374–2385 (2023)
10. K.W. Ferrara et al., Evaluation of tumor angiogenesis with US: imaging, doppler, and contrast agents. Acad. Radiol. **7**(10), 824–839 (2000). ISSN: 1076-6332. https://doi.org/10.1016/S1076-6332(00)80631-5. https://www.sciencedirect.com/science/article/pii/S1076633200806315
11. D. Fischerova, Ultrasound scanning of the pelvis and abdomen for staging of gynecological tumors: a review. Ultrasound Obstetr & Gynecol **38**(3), 246–266 (2011)
12. G. Peng et al., Automated 3D ultrasound image segmentation to aid breast cancer image interpretation. Ultrasonics **65**, 51–58 (2016)
13. K. Gupta et al., Ultrasound placental image texture analysis using artificial intelligence to predict hypertension in pregnancy. J. Mater.-Fetal & Neonatal Med. 1–8 (2021)
14. J.H. Lee et al., Deep learning-based computer-aided diagnosis system for localization and diagnosis of metastatic lymph nodes on ultrasound: a pilot study. Thyroid **28**(10), 1332–1338 (2018)
15. X. Liang et al., CR-unet-based ultrasonic follicle monitoring to reduce diameter variability and generate area automatically as a novel biomarker for follicular maturity. Ultrasound Med. & Biol. **46**(11), 3125–3134 (2020)
16. S. Liu et al., Deep learning in medical ultrasound analysis: a review. Engineering **5**(2), 261–275 (2019)
17. T. Liu et al., Automated detection and classification of thyroid nodules in ultrasound images using clinical-knowledge-guided convolutional neural networks. Med. Image Anal. **58**, 101555 (2019)

18. X. Liu et al., Deep learning for ultrasound localization microscopy. Trans. Med. Imag. **39**(10), 3064–3078 (2020)
19. U.-W. Lok et al., Fast super-resolution ultrasound microvessel imaging using spatiotemporal data with deep fully convolutional neural network. Phys. Med. & Biol. **66**(7), 075005 (2021)
20. Y.-J. Mao et al., Breast tumour classification using ultrasound elastography with machine learning: a systematic scoping review. Cancers **14**(2), 367 (2022)
21. L. Milecki et al., A deep learning framework for spatiotemporal ultrasound localization microscopy. Trans. Med. Imag. **40**(5), 1428–1437 (2021)
22. C.D.L. Nascimento et al., Breast tumor classification in ultrasound images using support vector machines and neural networks. Res. Biomed. Eng. **32**, 283–292 (2016)
23. S.H. Rossi et al., Imaging for the diagnosis and response assessment of renal tumours. World J. Urol. **36**, 1927–1942 (2018)
24. R.J.G. van Sloun, R. Cohen, Y.C. Eldar, Deep learning in ultrasound imaging. Proc. IEEE **108**(1), 11–29 (2020). https://doi.org/10.1109/JPROC.2019.2932116
25. R.J.G. van Sloun et al., Super-resolution ultrasound localization microscopy through deep learning. Trans. Med. Imag. **40**(3), 829–839 (2020)
26. Y. Wang, Q. Zeng, Ovarian tumor texture classification based on sparse auto-encoder network combined with multi-feature fusion and random forest in ultrasound image. J/ Med. Imag. Health Inf. **11**(2), 424–431 (2021)
27. W. Wei et al., A deep learning approach for 2D ultrasound and 3D CT/MR image registration in liver tumor ablation. Comput. Methods Progr. Biomed. **206**, 106117 (2021)
28. M. Xian et al., Automatic breast ultrasound image segmentation: a survey. Pattern Recogn. **79**, 340–355 (2018)
29. N. Yadav, R. Dass, J. Virmani, Objective assessment of segmentation models for thyroid ultrasound images. J. Ultrasound **26**(3), 673–685 (2023)
30. Z. Yu et al., A deep convolutional neural network-based framework for automatic fetal facial standard plane recognition. J. Biomed. Health Inf. **22**(3), 874–885 (2018). https://doi.org/10.1109/JBHI.2017.2705031

Appendix
Deep Learning and HPC for Image Processing

Nowadays, the use and applicability of *Artificial Intelligence* (AI) for signal processing, in particular images and videos, is widespread through the availability of large data sets and the development of hardware and parallel computing techniques. Developing efficient artificial intelligence methods for processing, approximation, and classification of images becomes relevant. Artificial intelligence methods take advantage of image redundancy, recognise complex patterns, and provide quantitative assessment for images (e.g., classification), showing high potential to assist the user in acquiring accurate and reproducible results. *Machine learning* (ML) is a branch of artificial intelligence that uses large data sets to create systems that learn or improve their performance. Through optimising a network of variables, ML methods allow us to learn a proper transformation from an input set to an output set. *Deep learning* is a subset of ML that processes and automatically learns mid-level and high-level abstract features from large data sets of images through multi-layered neural networks.

The area of AI in health and healthcare has seen rapid developments due to the increasing availability of health data, combined with unprecedented advances in AI and computation capabilities. AI systems can support humans, allowing them to optimise the diagnostic examination workflow and improve the analysis of diagnostic images. The most recent algorithmic progress enhances the quality of the acquired data and reduces the scan time; moreover, it can speed up the diagnosis and increase the diagnostic accuracy by supporting the physician in the analysis of the images (e.g., ultrasound and magnetic resonance) and the analysis of laboratory or histological data. Through quantitative information extracted by AI tools, radiologists might spot problems that are not immediately obvious to the human eye, especially in the case of advanced imaging techniques, such as functional imaging. AI, ML, and DL allow medical specialists to integrate imaging data with information from electronic medical records (e.g., longitudinal patient data, population health platforms) to provide a rich source of new insights for medical research, such as the identification of new protocols for imaging. Combining acquired data with predictive analysis, ML, and DL methods provides signifi-

S. Cammarasana and G. Patanè, *Real-Time Processing of Ultrasound Images*, Synthesis Lectures on Visual Computing: Computer Graphics, Animation, Computational Photography and Imaging, https://doi.org/10.1007/978-3-032-05805-8

cant benefits, supporting the management and prioritisation of medical analysis, improving intervention planning and preventing unnecessary re-admissions and further testing. From the technological perspective, all these aspects imply the need to expand their intra-operative capabilities, create operative equipment compatible with different imaging methodologies or the procedural environment, and develop devices/accessories for navigation and interaction.

We discuss main trends, methodologies, and technologies on deep learning (Sect. A.1), high-performance computing (Sect. A.2), and HPC resources for deep learning (Sect. A.3).

A.1 Deep Learning

Deep learning is a class of machine learning methods with multiple levels of representation, and it is composed of several non-linear layers that, starting with the raw input, transform the representation from the previous level into a representation at a higher level, with a reduction of the number of parameters that are applied to represent the image. Deep learning methods attain very complex features and inferences with the account of large data sets and the composition of a large set of layers. In this class of methods, one of the main distinctions is between image-to-label and image-to-image models: the goal of image-to-label models is to classify the training data according to two or more classes, for example, objects family or biomedical pathologies; the goal of image-to-image models is to start from an image and reconstruct another image as output with specific properties, as denoising or super-resolution problems. Deep learning methods are also classified into supervised and unsupervised, according to the prior knowledge of the output label/image.

Among deep learning methods, the convolutional operator is widespread for its property of local description and interpretation of images; its application leads to a sub-class of deep learning models named *Convolutional Neural Networks* (CNNs). The convolution operator captures essential features inside the images, such as contrast pixels and edges that are more representative of the objects in the image, reducing memory use and computational cost. Also, the convolution operator allows the sharing of the parameters (i.e., the model's variables) among different patches of the images; this property enables the network to account for the information redundancy inside each image.

Each layer of a convolutional network generally performs three steps (Fig. A.1). In the first step, the layer applies multiple convolutions through a set of operators (i.e., kernel filters). Next, the pooling function applies further changes to the layer's output according to the statistical values of nearby outputs. For example, *max pooling* modifies an output value according to the maximum value of its rectangular neighbourhood region. In contrast, *mean pooling* considers the average value of the neighbourhood region. The pooling operator generalises features extracted by convolutional filters and helps the network recognise the features independent of their location in the image, thus emphasising the convolutional aspect of CNNs, where neighbourhood values impact any given node. Finally, activation functions introduce a non-linear element into the network data processing, allowing each

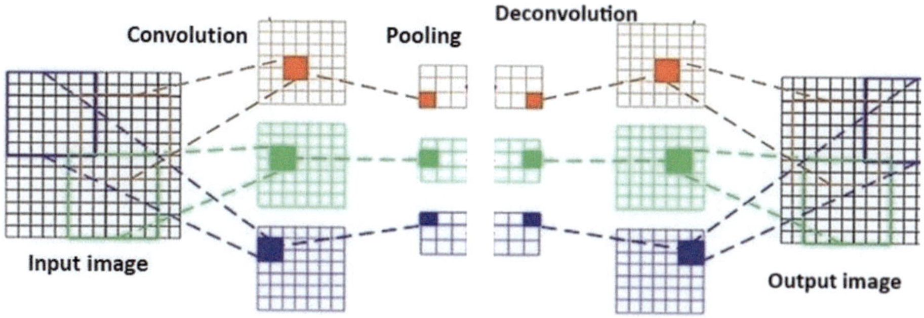

Fig. A.1 Example of convolutional neural network

layer to operate on the input data and feed the output information to the next layer. The role of an activation function is to produce a mapping from an input real number to a real number within a specific range to determine whether or not the information within the node is useful. The activation function's values depend on the activation function type used, e.g., sigmoid, rectified linear unit, and softmax.

Each kernel filter is associated with a set of variables (i.e., weights) optimised until the network can approximate the desired output properly. The number and dimension of each kernel depend on the architecture/purpose of the network, and these are two of the main hyperparameters to be optimised to improve the network's accuracy. Additional relevant hyperparameters are the number and connections of the layers (i.e., the network's depth) and the optimisation method's accuracy.

Given a network with L layers, a set of kernel filters F each of size $m \times n$, usually selected as odd numbers (e.g., 3×3), we apply the convolution at layer l of the filter f to the input $X(l)$ as

$$X_{i,j}(l+1) = \sum_{h=1}^{m} \sum_{k=1}^{n} f_{h-\lceil m/2 \rceil, k-\lceil n/2 \rceil} \cdot X_{i,j}(l).$$

Some relevant parameters are the padding, which allows the convolution operator to process boundary values, and the stride, which allows the operator to process all the data or a selection of it. In addition, the bias adds a variable as a linear component of the convolution filter to improve the approximation of linear transformations between consecutive layers. Finally, dropout regularisation is a computationally cheap way to regularise a deep neural network by probabilistically removing inputs to a layer and generating more robust networks.

We introduce forward propagation, providing the data to the first layer and advancing the network through the designed layers. The starting values of the variables are initialised through a proper method (e.g., normal distribution) and then optimised through an optimisation algorithm (e.g., Adam [9]). After the computation of the predicted output $\hat{y}$ of the network, this value is compared with the expected output y through a loss function, as mean squared error (Eq. (A.1)), or mean averaged error, or logarithmic-based error. With n input data samples, the loss function is defined as

$$\mathcal{L} = \frac{1}{n}\sum_{i=1}^{n} \|y_i - \hat{y}_i\|^2. \tag{A.1}$$

Backpropagation is one method of training a network by minimising the loss function, which represents the accuracy of the network to predict the desired output. To minimise the loss, a gradient descent optimisation algorithm [8] is applied, and the errors are back-propagated from the last up to the first layer of the network. Gradient descent computes each weight as the negative of the learning rate multiplied by the partial derivative of the loss function with respect to the weights. Backpropagation reduces the number of calculations required to compute the gradients compared to forward propagation and reduces the approximation error of the network.

Among image-to-image deep learning methods, several architectures are defined by the number of layers, their connections, and the setting of the hyperparameters. Autoencoders (Fig. A.1) are widespread for their efficiency. They are composed of two main steps: the encoding phase projects the information extracted by convolution operators into a smaller representation composed of a reduced number of elements in a higher-dimensional space; the decoding phase projects back to a dimensional space and resolution as the expected output data, which may or may not be the same as the input. The optimiser learns the reduced number of variables that codify the representation of the image from the encoded data to the decoded data. Furthermore, connections among layers of the decoding and encoding phase allow the network to increase the accuracy of the approximation.

Generative Adversarial Networks (GANs) efficiently represent the training data, making them more effective at generating predictions similar to the expected target of the data set (Fig. A.2). A GAN comprises two concurrent networks: a discriminator D and a generator G. The generator, for example, an autoencoder network, generates a prediction from an input image; the optimiser trains the network to generate a predicted output that resembles the expected output (i.e., the target of the data set) according to a defined loss. The discriminator is a network trained to classify targets and generated images; the network weights are optimised to distinguish the first from the second. The communication between the two

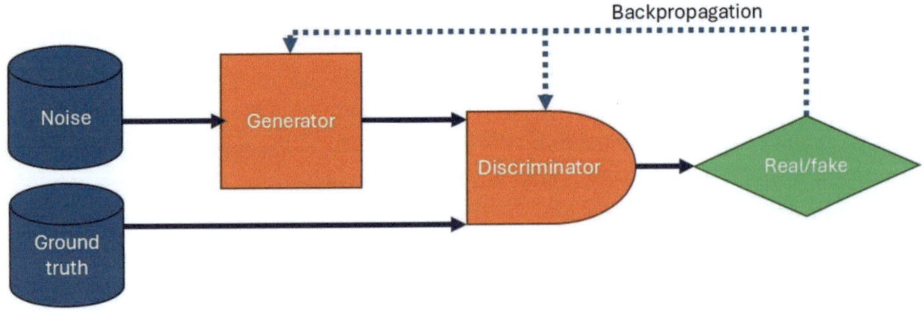

Fig. A.2 Example of generative adversarial network

networks is bidirectional and concurrent: the generator provides its output to the discriminator, which classifies it as a target or a generation. At the same time, the discriminator offers feedback to the generator as a quantification of the number of fake images that have been correctly classified as the output of the generator itself. Unlike standard CNNs, where the measure of the accuracy of the prediction is performed through a static indicator (e.g., the mean squared error), GANs analyse the accuracy of the prediction with respect to the target through an additional network, which increases the accuracy of the prediction with respect to the expected target. A complete survey on deep learning methodologies for image processing is presented in [6].

A.2 High-Performance Computing

Technological evolution allows us to acquire large amounts of data elaborated by complex multidisciplinary applications. Parallel computing enables us to solve big data problems exceeding the limits of memory and reducing the computation time through the synchronised usage of more processes. The method and the data are divided into multiple portions allocated to one or more processes and solved concurrently. Applying *High-Performance Computing* (HPC) methodologies, accounting for modern HPC resources (e.g., computing clusters, cloud computing), and developing proper parallel implementations of processing algorithms have become relevant to accelerate intensive computation for solving problems in numerous data science applications.

Modern HPC resources are typically highly heterogeneous systems composed of multiple CPUs and accelerators; the latest generation CPUs are hierarchical, accounting for multi-core processors and supported by powerful specialised accelerators, such as graphics processing units (GPUs). These resources offer significant computational power in the order of TFLOPS, which is achieved by exploiting such huge parallel architectures. The development of efficient software takes into account efficient data management; the computational performance is dominated by communications among processes and memory traffic rather than floating-point operations of the individual process; moreover, the different kinds of processors and accelerators pose new and significant issues in terms of memory management and data access.

Over the past decade, HPC has been increasingly introduced into image processing and biomedical applications to solve computational and significant data challenges. In this context, HPC is widespread for the training of learning models in US processing; for example, for the localisation of common carotid artery transverse section through RCNN [4], automatic segmentation of the carotid artery and internal jugular veins [2], foetal standard planes recognition [7], and segmentation and classification of anatomical structures [5]. HPC and cloud computing also pose new medical data privacy and security challenges. For a deeper analysis of HPC for biomedical data, we refer the reader to the DeepHealth project [1].

The main reasons for using a parallel environment are access to larger memory and higher computational power. Parallel computing can be divided into two main categories: data parallelism and model parallelism. In the first case, all the processes execute the same algorithm on different data. In contrast, in the second case, all the processes access the same data while executing different algorithm parts. The choice between the two approaches depends on the complexity of the data rather than the independence of the algorithm's subroutines. In both cases, communication and balancing among processes and data distribution become relevant for optimising the computational cost without facing bottleneck issues. Unfortunately, the memory and the overall computing power can not be calculated as the simple sum of the individual components due to the complexities arising from the communication between the processes and the sharing or duplication of the necessary data.

To measure the efficiency of a parallel implementation, we introduce some quantitative measures that account for the computational performance of single-process execution. Given p processes, and defining T_1 as the execution time of an algorithm on a single processor and T_p as the execution time on p processors, we define the *speedup* as $S_p = T_1/T_p$. In practical cases, $S_p \leq p$, since the execution time on p processes is higher than the execution time on one process divided by the number of processes, i.e., $T_p \leq T_1/p$. To quantify the *efficiency* of the parallelisation, we introduce a quantitative measure $E_p = S_p/p$ that accounts for the distance of the parallelisation from an ideal distribution of the algorithm among the processes; in fact, $E_p \leq 1$. Different reasons exist for the loss of efficiency in parallelisation: the use of various processes includes the addition of communication among them and synchronisation in the data communication that leads to barriers in the method and consequently to an increase in the execution time. Then, the load imbalance among processes implies that they have different amounts of work, leading to idle time for some processes. Finally, the algorithm has inherently sequential subroutines, and their execution times cannot be reduced independently of the number of processes adopted. Adding these overheads implies a loss of efficiency in the parallel code.

This result is summarised by Amdahl's Law (Fig. A.3). Given F_s be the sequential fraction of a code, and F_p the parallel fraction with p processes, with $F_p + F_s = 1$, the global execution time T_p is given by the execution time of the sequential part $F_s T_1$ plus the execution time of the parallel part $F_p(T_1/p)$, where T_1 is the execution time of one process; as result: $T_p = T_1(F_s + F_p/p)$. In addition, the parallelisation introduces a communication overhead among processes that further reduces the speed-up. Amdahl's Law is then refined as

$$T_p = T_1(F_s + F_p/P) + T_c,$$

where T_c is a fixed communication time and establishes that the number of processes should stay within the ratio of scalar execution time and communication overhead since the *efficiency* tends to decrease as the number of processes increases.

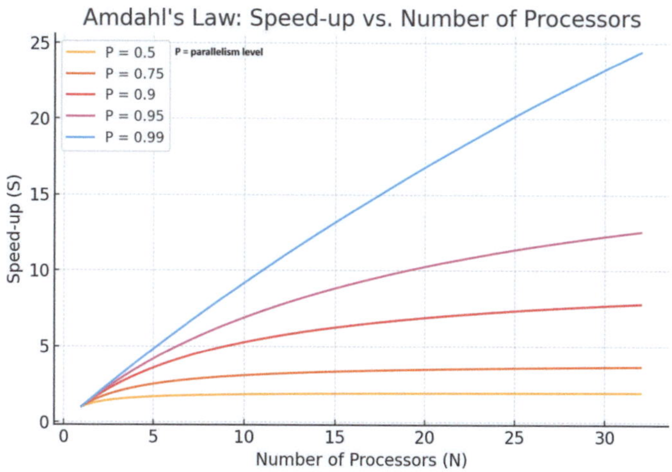

Fig. A.3 Amdahl's Law showing the speed-up (y-axis) with respect to the number of processes (x-axis) for different fractions of parallelisable code

A.3 HPC Resources for Deep Learning

HPC resources are usually composed of multiple nodes, each consisting of a set of multi-core CPUs and a set of GPUs. Each node shares the same private memory, while the passage of data between different nodes is managed through inter-node communication methods (Fig. A.4). The parallelism inside a single node is more efficient when the number of processes is not greater than the number of cores, since the communication among processes is faster, but the quantity of memory is limited. In contrast, multi-node parallelisation allows the HPC model to account for more memory and apply more processes. In this context, we can exploit HPC for deep learning as

- *data parallelism* by splitting the images among different processes;
- *model parallelism* by splitting the algebraic operations among different processes.

The *data parallelism* involves the distribution of the images among the nodes of the computing resource. This allows us to increase the batch size, i.e., the number of images loaded and processed simultaneously for each process. The *model parallelism* involves the parallelisation of the operations performed by the network layers. In particular, the convolution operation between the input data $\mathbf{X}$ and the convolution filter $\mathbf{K}$ is equivalent to multiplying the input data by a Toeplitz matrix. Given $\mathbf{X}$ as a $n \times n$ matrix, and $\mathbf{K}$ as a $m \times m$ matrix, with $n = 3$ and $m = 2$, we unroll $\mathbf{k}$ into a sparse matrix of size $(n - m + 1)^2 \times n^2$, and $\mathbf{X}$ into a $n^2 \times 1$ vector. We can compute the sparse matrix-vector multiplication as

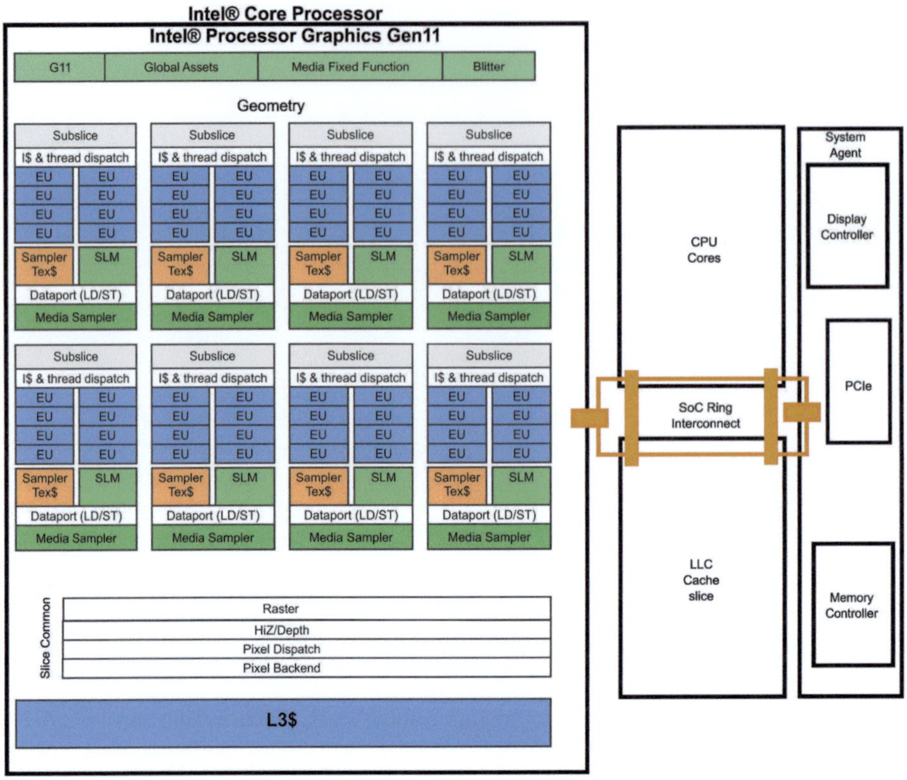

Fig. A.4 Example of intel processor graphics Gen11 SoC. Image courtesy of [3]

$$\begin{pmatrix} \mathbf{K}_{11} & \mathbf{K}_{12} & 0 & \mathbf{K}_{21} & \mathbf{K}_{22} & 0 & 0 & 0 & 0 \\ 0 & \mathbf{K}_{11} & \mathbf{K}_{12} & 0 & \mathbf{K}_{21} & \mathbf{K}_{22} & 0 & 0 & 0 \\ 0 & 0 & 0 & \mathbf{K}_{11} & \mathbf{K}_{12} & 0 & \mathbf{K}_{21} & \mathbf{K}_{22} & 0 \\ 0 & 0 & 0 & 0 & \mathbf{K}_{11} & \mathbf{K}_{12} & 0 & \mathbf{K}_{21} & \mathbf{K}_{22} \end{pmatrix} \begin{pmatrix} \mathbf{X}_{11} \\ \mathbf{X}_{12} \\ \mathbf{X}_{13} \\ \mathbf{X}_{21} \\ \mathbf{X}_{22} \\ \mathbf{X}_{23} \\ \mathbf{X}_{31} \\ \mathbf{X}_{32} \\ \mathbf{X}_{33} \end{pmatrix} = \begin{pmatrix} \mathbf{K}_{11}\mathbf{X}_{11} + \mathbf{K}_{12}\mathbf{X}_{12} + \mathbf{K}_{21}\mathbf{X}_{21} + \mathbf{K}_{22}\mathbf{X}22 \\ \mathbf{K}_{11}\mathbf{X}_{12} + \mathbf{K}_{12}\mathbf{X}_{13} + \mathbf{K}_{21}\mathbf{X}_{22} + \mathbf{K}_{22}\mathbf{X}23 \\ \mathbf{K}_{11}\mathbf{X}_{21} + \mathbf{K}_{12}\mathbf{X}_{22} + \mathbf{K}_{21}\mathbf{X}_{31} + \mathbf{K}_{22}\mathbf{X}32 \\ \mathbf{K}_{11}\mathbf{X}_{22} + \mathbf{K}_{12}\mathbf{X}_{23} + \mathbf{K}_{21}\mathbf{X}_{32} + \mathbf{K}_{22}\mathbf{X}33 \end{pmatrix}$$

and convert the result into a $n - m + 1$ square matrix

$$\begin{pmatrix} \mathbf{K}_{11}\mathbf{X}_{11} + \mathbf{K}_{12}\mathbf{X}_{12} + \mathbf{K}_{21}\mathbf{X}_{21} + \mathbf{K}_{22}\mathbf{X}22 & \mathbf{K}_{11}\mathbf{X}_{12} + \mathbf{K}_{12}\mathbf{X}_{13} + \mathbf{K}_{21}\mathbf{X}_{22} + \mathbf{K}_{22}\mathbf{X}23 \\ \mathbf{K}_{11}\mathbf{X}_{21} + \mathbf{K}_{12}\mathbf{X}_{22} + \mathbf{K}_{21}\mathbf{X}_{31} + \mathbf{K}_{22}\mathbf{X}32 & \mathbf{K}_{11}\mathbf{X}_{22} + \mathbf{K}_{12}\mathbf{X}_{23} + \mathbf{K}_{21}\mathbf{X}_{32} + \mathbf{K}_{22}\mathbf{X}33. \end{pmatrix}.$$

This operation has lower complexity than a fully general matrix-matrix multiplication and is parallel among the processes inside each node.

References

1. M. Caballero, J.A. Gomez, A. Bantouna, Deep-learning and hpc to boost biomedical applications for health (deephealth), in *32nd International Symposium on Computer-Based Medical Systems* (IEEE, 2019), pp. 150–155
2. L.A. Groves et al., Automatic segmentation of the carotid artery and internal jugular vein from 2D ultrasound images for 3D vascular reconstruction. Int. J. Comput. Ass. Radiol. Surg. **15**(11), 1835–1846 (2020)
3. Intel CPU, https://www.intel.com/content/www/us/en/developer/articles/technical/dpcpp-for-intel-processor-graphics-architecture.html. Accessed 06 Feb 2025
4. P.K. Jain et al., Localization of common carotid artery transverse section in B-mode ultrasound images using faster RCNN: a deep learning approach. Med. & Biolog. Eng. & Comput. **58**(3), 471–482 (2020)
5. T.R. Patel et al., Performance characterization of single and multi GPU training of U-Net architecture for medical image segmentation tasks, in *Proceedings of the Practice and Experience in Advanced Research Computing on Rise of the Machines (Learning)* (ACM, 2019)
6. W. Pedrycz, S.-M. Chen, *Deep Learning: Concepts and Architectures* (Springer, 2020)
7. B. Pu et al., Automatic fetal ultrasound standard plane recognition based on deep learning and IIoT. IEEE Trans. Ind. Inf. **17**(11), 7771–7780 (2021)
8. S. Ruder, An overview of gradient descent optimization algorithms (2016). arXiv:1609.04747
9. Z. Zhang, Improved Adam optimizer for deep neural networks, in *26th International Symposium on Quality of Service* (IEEE, 2018), pp. 1–2

MIX
Papier aus verantwortungsvollen Quellen
Paper from responsible sources
FSC® C105338

If you have any concerns about our products,
you can contact us on
ProductSafety@springernature.com

In case Publisher is established outside the EU,
the EU authorized representative is:
Springer Nature Customer Service Center GmbH
Europaplatz 3, 69115 Heidelberg, Germany

Printed by Libri Plureos GmbH
in Hamburg, Germany